Seth Green

Trout culture

Seth Green

Trout culture

ISBN/EAN: 9783337143220

Printed in Europe, USA, Canada, Australia, Japan

Cover: Foto ©berggeist007 / pixelio.de

More available books at **www.hansebooks.com**

BY

SETH GREEN.

PUBLISHED BY SETH GREEN AND A. S. COLLINS,

CALEDONIA, N. Y.

ROCHESTER, N. Y.;
PRESS OF CURTIS, MOREY & CO., UNION AND ADVERTISER OFFICE.
1870.

PREFACE.

Having been one of the first to practice Fish-Culture in this country, and being now, perhaps, the largest and most successful Trout-Culturist in America, so many requests have been made to me "to write a book," that I have at last concluded to try it. This little treatise is intended as a manual for Trout-Culturists, and is written especially for those who wish to RAISE Trout. It deals only with the practical part of the business. No scientific terms have been used, when other and more common names would equally convey the meaning. Neither has any scientific enquiry been entertained, any further than it would be of practical use to the fish-farmer. The science of Fish-Culture is yet in its infancy; and I claim only for this book, that it is an accurate description of my own method and practice. Still it should be remembered that this practice is the result of some years of study, labor and numberless experiments. I am very well convinced that I can handle the rod better than the pen, and beg that the book shall be judged by its matter rather than by its language and style.

I am under obligation to my partner, Mr. A. S. Collins, for assistance in preparing this work.

SETH GREEN.

Golden Rule for Pisciculturists.

"Never put off 'till to-morrow that which you can do to-day."

CONTENTS.

INDEX.

TROUT CULTURE.

CHAPTER I.

INTRODUCTORY REMARKS.

It is only lately that the culture of fish has claimed the
attention of our people. The importance of the art has
long been recognized abroad, but in our own country the
government has been slow to move in the matter, although
the supply of fish food is daily decreasing. Still individual
enterprise here will need no aid from government, except
in the matter of stocking and protecting streams which an
individual cannot control. The importance of fish as food,
and their cheapness, render it a matter of great importance
that the supply shall not be diminished, but very much increased. This can be done in no way so well and so quickly
as by artificial breeding. By this means, fish can be raised
as well as any other stock, and made so cheap that the
poorest in our land can have abundance of good and nour-
ishing food. I do not refer to Trout alone. The attention
of individuals working for profit has naturally been given
to the highest priced fish; but the experience obtained in
Trout raising will lead, and is even now leading, the way
to the production of all other kinds. The time will come
and come soon, when our rivers, lakes and streams will be
abundantly stocked with those fish to which they are best

2

adapted. For the lakes, the Whitefish, Salmon Trout, Herring, Black Bass and Wall-eyed Pike; for the rivers, the Yellow Pike, Black Bass, Shad and Salmon; for still and deep streams, the Bullhead or Catfish, the Perch and many other kinds of coarse fish; for the swift mountain stream, the Trout. Even Gold Fish, which are a good coarse pan fish, can be made to abound in all our rivers and bays. It has been proved beyond a doubt, that with very little care and expense these fish can be made to abound in all our waters. But this requires government aid, since individuals owning parts of streams will not hatch out fish there at their own expense for the benefit of all other owners of the stream. Besides this, special legislation seems to be required to get fish-passes constructed over the numerous dams in our rivers and to prevent substances destructive to the fish being thrown into our streams, such as saw-dust and the refuse of paper mills, tanneries, dyeing establishments, &c., &c.

But if with comparatively little care and expense our great rivers can be stocked, in the meanwhile there is room enough for private enterprise. There are few farmers in our country who do not have upon their land a lake, or spring, or clear running stream. If these men knew how easily they could turn this water to profit, not only by raising food supply for themselves, but a supply for the city and village market, there would soon be very few waters without their finny inhabitants. How much this would add to the wealth of the country any one can see at a glance. Of course this art, like any other, demands study and practice. But we do not hesitate to say that at the present time an acre of water can be made to pay far more than an acre of land. It is with difficulty that I refrain from saying more about the importance of fish culture in general. It is an art in which I am deeply interested. But as this book has a special subject, it would perhaps be out of place to say more about it now.

Trout raising is that branch of Fish Culture most generally pursued at the present time. There are several reasons for this—it is the best understood, one of the easiest to practice, and the fish bring the highest price. In this treatise we will try to commence where a person who wishes to raise Trout would naturally require information. The situation and laying out of Ponds will first be described, and as those who start in the business generally buy or otherwise obtain impregnated eggs, they will be told what to do with the eggs, then how to keep the young Trout raised from them. Afterward we will speak of the adult Trout and of the process of taking eggs from them for other crops.

This book gives only a description of the method of raising Trout, but Salmon, Salmon Trout and Whitefish may be hatched in the same way and with the same apparatus. Perhaps at some future time the author may give to the public a book on the culture of other Fish.

CHAPTER II.

TROUT PONDS.

LOCATION.—It is very easy with good spring water to raise a *few* trout anywhere in temperate latitudes. But to raise a large number requires care in the selection of a location. Plenty of pure spring water is the first and most essential requisite. The spring, or one of the springs, if there are several, should have a fall of two or three feet, and a fall of five to ten feet of the whole volume of water is decidedly advantageous. If the supply of water is very large, it diminishes the necessity of a fall. The water from a spring remains (near its source) at nearly the same temperature during the whole year, and is the best for Trout raising. The water from a brook which does not rise higher than sixty-five degrees in summer, may be used to supply ponds for adult Trout; but spring water is absolutely necessary for hatching purposes. It is not a good plan to dam up a stream which varies in volume, and so make ponds. There should be enough level land by the side of such a stream to make ponds supplied by the stream; and it is best to have a stream much greater in volume than is necessary for the ponds, so that there will always be a good supply of water, and will be no trouble with the surplus water after a freshet. The reasons for these and other essentials will, it is hoped, be found in the following pages. A good knowledge of the whole system of Trout Culture is essential in choosing the very best location. It is best to have your ponds near your house, or have a man in charge living at the ponds. Of course your Trout may never be molested, but "an ounce of prevention is worth a pound of cure."

LAYING OUT PONDS.—The diagram on page 14, represents a series of ponds, in all of which the same water is used. This plan is generally considered the best, for several reasons. It economizes the water and space, and is most convenient for changing the fish from one pond to another. It is not necessary that the ponds should be in a straight line. Where the location demands it they may be turned so as to lie in a direction nearly or quite parallel with one another. This is easily done by bending the raceways, and lengthening them if necessary, only a curved raceway is sometimes not so convenient as if it was straight. The sides of the ponds may be walled up with stones, laid without mortar, unless the soil is very sandy. Wood may be better for the sides and bottoms, but we are inclined to think is not worth its expense. If the sides of the ponds are laid up with mortar, let it dry thoroughly before letting the water in; then let the water run through it two or three weeks, or just long enough to purify the pond before putting any fish in it. It is as well to test it by putting in only a few fish at first; if the pond is not thoroughly purified the fish in it will turn blind. Ponds should not be built where much surface drainage will run into them; if they are so exposed the surface water should be carried off by a ditch. The Second and Third Ponds should receive an additional supply of water. The reason for this will be given further on. A general idea of the form and size of ponds can be gathered from the diagram without further explanation. If the supply of water is small, it is best to have as much fall between the ponds as the nature of the ground will allow. This fall aerates the water and makes it as good as new.

SHAPE OF PONDS, &c.—Where the supply of water is large it matters very little about the shape of ponds. The

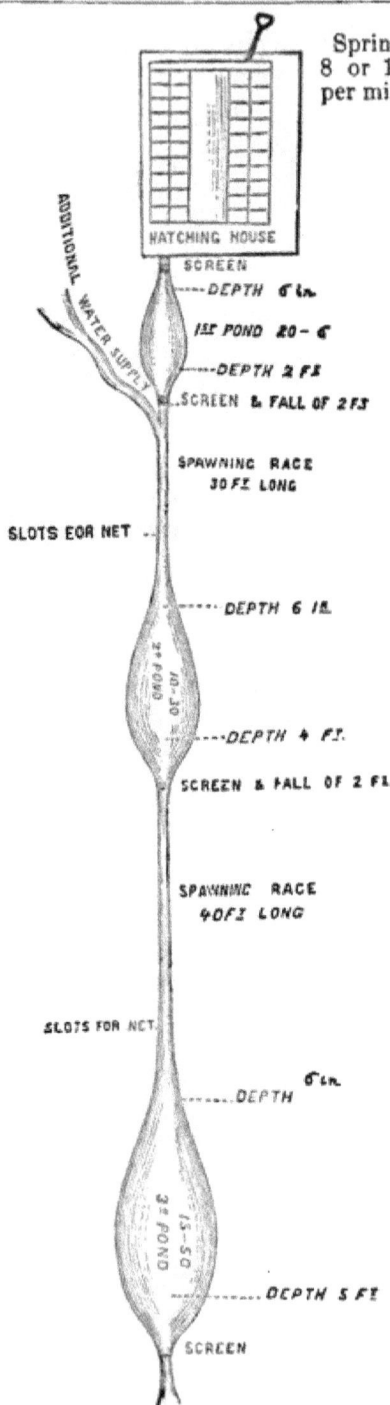

Spring running
8 or 10 gallons
per minute.

HATCHING HOUSE

SCREEN

ADDITIONAL WATER SUPPLY

----DEPTH 6 in.

1ST POND 20-6

----DEPTH 2 FT

SCREEN & FALL OF 2 FT

SPAWNING RACE
30 FT LONG

SLOTS FOR NET

----DEPTH 6 IN.

2ND POND 40-30

----DEPTH 4 FT.

SCREEN & FALL OF 2 FT

SPAWNING RACE
40 FT LONG

SLOTS FOR NET

----DEPTH 6 in.

3RD POND 75-50

----DEPTH 5 FT

SCREEN

In this diagram the proportions are not preserved, but as the dimensions of each part are given, it will not mislead. It is only intended to give a general idea of the construction of a series of ponds.

best shape we believe to be the pear-shape, figured in the plate ; such a shape combining an equable flow of water in all directions and the greatest amount of surface, with the least difference in the temperature of the water. If the nature of the ground demands other shapes, the ponds should be made long, narrow and deep, rather than broad and shallow. The depth of the pond is indicated in the plate, and will answer for any size of ponds. It is better for any one wishing to raise a large number of fish, to have several series of ponds, than to attempt raising a larger number by increasing the size of the ponds. Fish do not feed so well in large ponds, are not so easily taken care of, and eat each other more.

RACEWAYS.—The Second and Third ponds should have a long, narrow raceway where the water enters—about thirty or forty feet long, four feet wide and six inches deep. The sides of the raceway should be made of one and a half inch plank, one foot in width. This will answer for both natural and artificial impregnation. The raceway is required not only for the purpose of spawning, but as a resort for the fish at all seasons of the year. Fish will go into this shallow graveled race, into the quick running water, to free themselves from the parasites which often trouble them ; or they will go there if they are out of health and condition from any cause. This raceway must be filled with coarse gravel, and the bottom of the pond made to slope gently up to the raceway.

The head of the raceway is to be carefully looked after. If a series of ponds are made, then the screens between will keep the fish from running one to the other ; but if single ponds are used, each supplied with separate water from a stream, then much attention must be paid to the screens where the water enters. It would be well if the water was brought into the pond through a long box, as

the water will very soon work around or under a short
box, and allow the fish to escape.   If the water enters with
a fall, it may be allowed to pour over upon an apron, constructed of thin slats, one-half or one-quarter of an inch
apart, and set edgeways.   This will let the water through
and keep the fish from running up.   Trout will run up
stream very freely, working their way through a small
passage, but will not try very much to run down stream.

BOTTOM OF PONDS.—It matters very little of what material the bottom is composed.   Anything—mud, clay or
moss is good, except gravel, and this is bad, not from the nature of the substance, but because the fish will spawn on
it and the eggs be lost.   Sometimes a person will wish to
construct a pond in a place where there are springs, or to
dam up the water and make a pond in a springy place.
Under such circumstances it is a good plan to fill the bottom entirely with gravel, as the fish would spawn there in
any case.   For such a pond make the borders very shallow, so that the little fish may run up into the shallow
water and escape from the large fish ; or have the pond so
arranged that after the fish have spawned they may be
removed.   Thus the eggs will hatch out and the little ones
grow without danger.   When the next season of spawning
comes the little fish may be removed into another pond
and the old ones let in again to spawn.   Such a pond is
good for any one wishing his establishment to run itself, as
with a little care he can raise many fish in it without much
trouble.   Very often the bottom of a pond is porus and
absorbs the water as fast as it runs in, so that there is
hardly any running from the proper outlet.   If you are
short of water and wish to use all you can get for another
pond, it is best to cement the bottom.   If you have no further use for the water, it makes no difference how it goes
off, that is if there are no holes in the bottom large enough

to let the fish escape, and the water keeps up to its proper level. Weeds or mosses of any sort are not necessary at the bottom, and if the supply of water is not large they will speedily become a nuisance. The quantity of Trout food which they will produce is of no account in an artificial pond where large numbers of Trout are kept, and they tend to foul the water by hiding dead fish, bits of meat, &c. &c. It is best, if possible, to have ponds so arranged that they can be entirely drained. This is necessary, sometimes, for cleaning or repairing the ponds, changing the fish from one pond in to another, &c. If the slope of the ground is sufficient to permit of such an arrangement, it will often save much labor in pumping or bailing. The drain pipe may be of pump logs, tile or pipe of any kind, and should be fixed in the lowest part of the bottom, or as near it as the level of the ground will allow. Still better would be a regular flume reaching from the bottom of the pond to the top. A bulkhead may be put in to raise the water as high as may be required, and a screen the whole size of the flume set in front. This large screen would be an additional advantage, as the larger the screen the less liable it is to clog up with leaves, moss, &c., and the greater will be the volume of water passing through.

SCREENS.—Screens may be made of common wire painted or tarred, of copper wire, or of galvanized iron wire. The last is the best, as it will last longest in proportion to its cost. The screens for keeping the small fry should be of fourteen threads to the inch ; for one year old fish five or six threads to the inch ; for two year olds four threads to the inch, and for three year olds three threads to the inch. Incline the screens to an angle of forty-five degrees, the top being farthest down stream. By inclining the screens in this manner a greater surface is exposed to the water than if they were placed perpendicularly. The sockets

should be made so that the screens will fit tightly and yet
be easily taken out to clean.

A very good screen for two and three year olds can be
made from strips of lath planed and nailed to a strong
frame, with quarter-inch openings between them.

WATER SUPPLY.—It is immaterial what kind of water is
used, whether hard or soft. Neither will so-called "min-
eral water " hurt the Trout unless the water is very strongly
impregnated. Trout have been known to live and thrive
in a stream one-sixth of whose volume was supplied by a
strong sulphur spring. Still the purest water is the best.
The essentials are that the stream shall be reasonably pure,
the volume of water nearly uniform, or so arranged that
the supply taken from it is uniform and the temperature
between thirty-six and sixty-five degrees.

The supply of water necessary for a given number of
Trout is yet unsettled. For a series of ponds turning out
one thousand large fish yearly, the water supply should fill
a four-inch pipe. This question will be treated more at
length in Chapter VI.

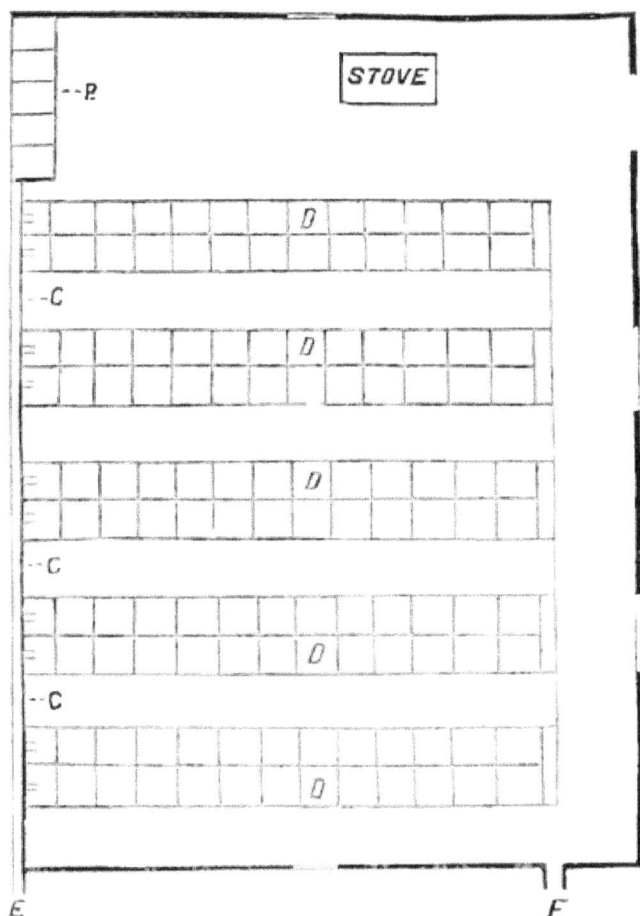

PLAN OF A HATCHING HOUSE.

Size 24 by 32 feet.   Scale $\frac{1}{8}$ inch to a foot.

A.—Inlet of Spring, 5 square inches.
B.—Filter, 6 feet by $1\frac{1}{2}$x$1\frac{1}{2}$.
C.—Distributing Trough, 5x5 inches.
D.—Troughs, 20 feet by 18 inches.
E.—Outlet of surplus water from Distributing Trough.
F.—Outlet of water from Hatching Troughs.

CHAPTER III.

Hatching House.

Size and Make.—If only a few eggs are to be hatched (say eight or ten thousand) no hatching house is necessary. The troughs may be placed in the open air, in any convenient place, and covered with a wire screen to keep out rats, mice and ducks. A light board cover must then be laid over them to shed the rain and snow and keep the eggs from exposure to the sunlight. A hatching house is much more comfortable to work in. A stove may be put in it and a fire started occasionally for warming one's fingers, &c., but it is not needed for hatching purposes, as spring water in these latitudes is warm enough. In our hatching house at Caledonia we run all the waste water into a large tank under the house, and as our water stands between forty-five and fifty degrees, even in the coldest weather, we find that the hatching house is a very comfortable place to work in without fire, though the thermometer outside points to zero. The house may be constructed of rough boards, or as expensive as you choose, but care should be taken to have a water-tight roof, as drops of water leaking through and falling into the troughs will kill the eggs underneath. Its size must be regulated by the number and extent of the troughs.

The windows in a hatching house should be few in number and provided with curtains or shutters, as the sun shining upon the spawn will kill it. We do not mean that a few minutes exposure to the rays of the sun will hurt the eggs, but a few hours exposure certainly will. Perhaps it would be well to have the windows, if possible, made on the north side of the hatching house, into which the sun will not shine in the winter season. Keep the hatching house clean. In fact, cleanliness is one of the cardinal virtues to the Trout raiser. He should have a clean house, should work with clean hands, and have all his. pans, spoons and utensils of every sort free from grease and dirt.

TROUGHS.—These should be made of seasoned timber, one and a half inches thick. They should be six inches deep and about fifteen inches wide, inside measurement. It would be better, perhaps, if the troughs were eight or nine inches deep, because then the water could be raised higher over the young Trout after they are hatched out. The difficulty in making them so deep is that when the sides of the trough are made so wide they are apt to warp or stretch apart at the top, and must be stayed in some way ; for instance, by strips nailed across. But the cleaner the trough is of all strips, elbows or grooves the better. The troughs are divided into squares or nests by cross strips set on the bottom at intervals of eighteen inches. The reason for this division into nests and for these cross strips will be seen further on. These strips may be made of half-inch stuff and cut two inches in width. There is no necessity for nailing them to the bottom ; fit them in accurately and set them edgeways at intervals of eighteen inches. As they do not need to be removed often, it is better to make them fit tightly. Other strips of the same stuff must be provided, to fit upon these and made wide enough to raise the water within a half inch of the top of the trough. As these need to be often moved, they must be made loose enough to take out, and yet fit accurately enough to raise the water over them when they are put in. A groove is sometimes made in which to run the strips, or shoulders nailed to the sides against which to set them. We do not recommend either of these, as it interferes with the equable flow of the water. New wood under the action of water develops a slimy sap, therefore it would be well to line the bottom and sides of the trough with glass. The troughs should have an inclination of about one inch in eight feet—just enough to let the water ripple gently over the cross strips. They should not be longer than twenty feet, or the air on the water will be exhausted before the water reaches the end of the trough. There is more danger of this after the eggs are hatched out and the troughs are full of young fish. This method is that which we pursue and have found most successful in practice. There are many other methods, all of course depending upon the same general principles, but we give only that one which we consider best. If possible the hatching house should be so far below the level of the spring from which its supply of water is derived, as to allow the troughs to be raised two or three feet from the floor. Where a large number of eggs are to be hatched, the inconvenience of stooping to care for them is very great.

WATER SUPPLY.—From the filter the water runs into the distributing trough or pipe, which runs along the head of all the hatching troughs. The water may be let into the hatching troughs by faucets, or through holes cut into the trough. These holes should be covered with netting, or the young fish will run up out of the troughs into the filter, or coarse gravel may be heaped up at the head of the trough through which the water will run, but through which the young fish cannot work their way. The supply of water for one trough should be equal to that coming through a half-inch hole with three inches head; just enough to make a gentle ripple over the cross-pieces. Be careful to get the troughs level crossways, and the strips true, so that when the water is running it will form an equal current over every part of each strip along the whole length of the trough. If the water runs unevenly the eggs will be washed into a heap, and many of them spoiled for lack of a proper circulation of water around them. This supply of water will be sufficient until the eggs are hatched out, when a somewhat larger supply can be allowed. The water should be brought directly from the spring in a pipe of some kind, in order to preserve the proper temperature and keep the water as free from sediment as possible ; and for the same reason the spring should be walled up to its smallest possible dimensions. If any surface water naturally runs into the spring, a ditch should be dug around the spring to lead it off. If the muddy surface water is suffered to run into the spring which supplies the troughs, the screens will very soon be choked up, and the sediment will find its way into the troughs in spite of all precautions and destroy the eggs.

FILTER.—The filter is a box six feet long by one and a half feet wide and one and one-half feet deep ; in which four or five flannel screens can be placed through which to filter the water before it passes into the troughs. The coarsest and cheapest red flannel is the best. It will rot and must be renewed once or twice in a season. Red flannel will last twice as long as any other. The flannel should be tacked on to frames running in grooves set at an angle of forty-five degrees, (the top down stream) so as to expose as much surface as possible to the water. If the hatching house is small, the filter may be placed outside, but is better under cover. If the spring is well protected the screens will not need cleaning very frequently. They should be cleaned as soon as they look dirty, however often that may be, and can be cleaned best by being taken out and washed with a soft brush.

Sediment destroys the eggs by suffocating them. Gravel
is put into the troughs in order to support the eggs at the
fewest possible points, and let the water touch as much of
the surface of the egg as possible. Sediment falling on the
egg keeps the water off and destroys its life as effectually
as being buried in the ground would destroy a man's life.
If sediment falls upon the eggs it may. be removed by
gently agitating the eggs with a feather, or better still,
by creating a current in the water with a feather, which
current the eggs will follow, and as they roll over, the sediment will drop off. But the Trout breeder has no business
to be troubled in this way. If his apparatus is constructed
right, and his filter properly attended to, there will not be
sediment enough in the troughs to hurt the eggs, from the
time they are put in until the fish are hatched out. The
pipe which is let into the spring should have wire netting
around it where the water comes in, to keep out impuri-
ties. This netting should be spread out so as to give a
greater surface than the mouth of the pipe. If the netting covers only the mouth of the pipe, every speck of dirt
which lodges on the netting diminishes by so much the
supply of water ; but if the surface of the netting is increased, much of it may be stopped up with dust without
lessening the supply of water. The best way is to take a
board, say one foot square for each inch of diameter of the
pipe, and run the pipe through a hole in the middle of the
board, fitting it well ; then nail netting on the edges of
the board, so as to bulge it out in a half globe form. This
should be looked after occasionally, but if the spring is
closely walled up, and the netting placed beneath the surface of the water, it will not probably need cleaning
through the season.

GRAVEL FOR TROUGHS.—The gravel for the troughs
should be quite fine—about the size of peas. It is better

to have it of a uniform size. Any kind of gravel is good
which is free from iron rust, as that kills the fish. If the
gravel is of some dark tint, the dead eggs, which turn milk
white, will show very plainly upon it, and may easily be
picked out. The gravel should be well washed before use,
and we would even recommend boiling it, to destroy any
eggs of insects which may be adhering to it. After the
nests are lined with glass the gravel may be put in, one
and one-half inches deep, which will bring it within one-
half inch of the top of the cross-piece.

IMPLEMENTS.—The implements of the fish-culturist are
few and simple. A few feathers may be kept on hand to
use in spreading the eggs when placing them in the troughs,
in collecting them for packing, and moving them in the
search after dead eggs. Several plans are in use for removing dead eggs from the trough. Some use a siphon to
draw them up; others bend wire into the shape of a small
spoon, or bend an eye upon the wire just large enough to
hold the egg. We recommend the use of nippers. These
may be made of wire bent into the shape of the letter U,
and flattened at the ends so that the extremities may be
about the eighth of an inch wide; round off the corners.
The length of the nippers should be 6 or 8 inches. A better one may be made of double wire, same shape, with a
small loop in each end. These will hold the egg without
trouble. A small homœopathic phial is used to examine the
eggs. The manner of its use is to fill it with water, put in
the eggs to be examined, cork it, hold it up before the win-
dow in a horizontal position, and with your microscope
look up through the side of the phial. This brings the
egg which lies at the bottom of the glass within the focus
of the microscope, and the water does not distort its shape.
This seems to be a very simple thing, and hardly worth
telling, but of the hundreds who have tried to examine
3

eggs in our hatching house, not a half dozen got it right
until told how to do it. The microscope need not be very
strong ; one magnifying eight or ten diameters is amply
sufficient. A small net will be of use in removing the
young fish from the troughs ; it should be about 6 inches
in diameter, in the shape of the letter D, with the handle
on the middle of the bend. It is very easily made by bending a wire in the desired shape, and twisting the two ends
together for a handle. Thin gauze of some kind should be
spread over the wire so tightly that the middle of the net
shall hang only a half inch below the level. An iron spoon,
well tinned or silvered, is used to remove the eggs. Some
six-quart tin milk-pans will be necessary, for a variety of
purposes. Eggs may be counted most easily by measuring
them. For this purpose take any small glass, such as a
very small tumbler, for instance, count out 500 or a 1,000
eggs, and with a file make a mark upon the glass as high
as they reach, and the measure is always ready to your
hand.

Chapter IV.

Treatment of Eggs.

Placing Eggs in the Troughs.—The eggs of a Trout are about one-sixth of an inch in diameter, and nearly round. They are generally of a light straw or salmon color. The color varies with the meat of the fish. The redder the meat, the more orange colored are the eggs. They are generally of a light yellow or amber color at first, and grow darker as the egg grows older. Their specific gravity is a little greater than that of water, so that they will sink in water, but may be easily moved in it. Suppose the eggs to be obtained and that you have them in a shallow pan. The water in the troughs should be raised by placing a narrow strip across the trough upon one of the two inch strips dividing the nests. (See pages 19 & 21.) Then sink the pan gently to the edge in the water of the trough, at the same time tipping the pan, so that the water in the trough and in the pan shall come together with as little current as possible. Then the edge of the pan may be sunk into the water, and by tipping the pan a little more, the eggs will flow out without injury By moving the pan while the eggs are running out, they may be spread uniformly over the bottom. If they fall in a heap, take the bearded end of a feather, and move the water with it in the direction you wish the eggs to go, and they will follow the current thus created. This may be done without touching the eggs with the feather. Distribute the eggs as evenly as possible over the surface of the nest.

The strip which was placed across the trough to raise the water should then be removed. Care must be taken that it be not removed so suddenly as to cause a rush of water, which would carry most of the eggs away with it. Raise the strip a little way from the bottom so as to let the water run out gradually, and when it is very nearly or altogether at the proper level, the strip may be removed entirely. Those who have a nursery attached to the troughs place the earliest eggs in the lower end of the trough, and keep placing them toward the top, so that the fish which are first hatched can run first into the nursery without disturbing the others. We practice placing the eggs in the highest end of the trough first, because the eggs earliest placed, hatch out first, and the water should be raised over them. If these first should be placed at the lower end of the trough, in order to do this the water must be raised over all the eggs, if at the beginning, by placing strips upon the nests in succession as the eggs hatch out, the water is left running upon the the unhatched eggs as usual. About five hundred may be placed in each nest eighteen inches by fifteen inches. This requires a word of explanation. Ten thousand or even twenty thousand, could be placed in a nest of the same size and would hatch out. In our own establishment we place as many as ten or fifteen thousand in each square. But as we wish the fish to live in the troughs a few months after hatching, we must reduce the number, and by taking out the eggs for sale we leave only five hundred or one thousand in each square. The philosophy of the thing is simply that the eggs require much less oxygen than the fish.

If the eggs are received from a Trout breeder, they should be left as received until the troughs are ready for them. It has sometimes occurred that the persons to whom we sent eggs, took the tin boxes out of the saw-dust in which they were packed, and set them in the water of their troughs, with the idea perhaps of getting the eggs in the box to the same temperature as the water in the troughs before un-packing them. This will surely kill the eggs in a few hours. Leave them in the original package until a few hours before you are ready to place them in the troughs. Then take out the tins and set them over or near the troughs, which will reduce or raise the temperature enough. Then empty the box into a tin pan full of water taken from the trough, pick out as much moss as you can readily with your fingers or nippers, and wash off the nest in the man-ner shown in directions for washing eggs in chapter VII.

If the eggs have had decent treatment on the way, that is not thrown about roughly or set near a red hot stove, you should find very few dead eggs in the boxes, not more than ten or twelve in one thousand. Should the eggs be found, on opening the box, run together in lumps instead of being evenly distributed, and turned to a dead white or milky color, it shows rough usage on the way, and does not necessarily show that the eggs were not impregnated or in good order when sent.

TEMPERATURE OF WATER AND TIME OF INCUBATION.—The length of time required to hatch out the eggs depends upon the temperature of the water. A general rule sufficiently accurate for all practical purposes is this: At fifty degrees the eggs will hatch out in fifty days, each degree colder takes five days longer, and each degree warmer five days less. The difference however increasing as the temperature falls and diminishing as it rises. The best temperature for hatching is between forty-five and fifty-three degrees. We are inclined to believe that the fish hatched at a temper-ature of about forty-five degrees and taking from seventy to seventy-five days to hatch, are stronger and longer lived, than those hatched in fifty days at fifty degrees. It may be well, also, to note that the eggs earliest taken will produce the best fish. The water of a spring can be reduced in temperature in winter by letting it run for a short distance exposed to the open air.

GROWTH OF EGGS, &c.—About the twentieth day, the young fish can be plainly observed in the egg. Put a few eggs in a small vial and with a magnifying glass the formation of the fish can easily be seen. Fish farmers send the eggs away at this time. Some of the eggs are not impregnated and at this stage of growth may easily be distinguished from the others as no fish forms in them. The dead eggs will turn to a milk or pearl white color, and should be removed with the nippers as fast as they are discovered. If left in the trough a fungus growth forms upon them which extends to other eggs in the immediate vicinity and kills them. Care should be taken in using nippers, not to hurt the other eggs; a very slight blow or jam from the nippers will be sufficient to destroy their vitality. Rats and mice in the hatching house often destroy many eggs; they are very fond of them and going into the troughs to get them they will destroy with their feet many more than they eat. A wire screen laid over the troughs will keep them out, but it is a much cheaper and just as affectual a way to keep them down by traps or poison. The eggs should be feathered over occasionally so that their whole surface may be exposed to the action of the water.

TRANSPORTATION OF EGGS.—We pack eggs in round tin boxes, about three inches wide and two and one half inches deep; a few small holes are punched in the bottom to let the water run off, as water left in the box will kill the eggs. Specimens of eggs from different parts of the square are first examined with the microscope to see if a good per centage is impregnated. If they are all right a six quart pan is filled with water to the height of the box in which the eggs are to be packed. The bottom of the box is then filled with moss, and the box placed in the pan and filled with water. The moss which we use is that which grows on stones and timbers, in wet places, such as the stones in a brook, or the timbers of an old dam. It may be collected and kept all winter in a damp place in the hatching-house. The bottom of the tin is filled with a piece of this moss, (roots downward,) somewhat depressed in the middle, so that the eggs shall not touch the sides of the box, the moss having previously been well washed to free it from dirt and insects. The rest of the moss to be used in packing must undergo a little more preparation. The green fibres must be cut with a pair of scissors from the roots. Only the green, soft and living fibres are used, and the roots, stems and dead leaves thrown away as useless. This fine moss must then be washed thoroughly. A very convenient way is to nail wire netting over the open bottom of an old soap box. Cut the moss into this, and dipping it into water, wash thoroughly. By simply lifting your box out of the water, you drain the moss. The eggs are then taken out of the trough, by being brushed with a feather into a spoon. If you wish to number them, fill your glass measure with water, and turn the contents of the spoon into it. When the five hundred or thousand eggs are measured, pour them into a ladle (small enough to go inside of the packing box), having previously filled the ladle with water; then sink the ladle beneath the water in the packing box, and by gently tipping and shaking it the eggs will fall to the bottom of the box, where they may be spread evenly over the moss with a feather. A layer of prepared moss must then be lightly laid over the eggs, (don't take the box out of the water,) and another five hundred or thousand eggs put in. Then fill the box with same kind of moss, take it out of the water, and leave it a little while, so that the water may drain off through the holes in the bottom, and the damp, spongy moss be left, an elastic and life-giving cushion to keep the eggs from feeling sudden jolts on the journey, and to supply them with oxgen. When the water is all drained off, the covers are to be placed on the boxes, and tied on with pack thread. If in any of these operations the box of eggs should fall out of your hands to the floor, it would probably kill nearly every egg. The tin boxes are to be packed in saw dust in a box or tin pail, the saw dust being first very slightly dampened. We generally pack our eggs in a tin pail, so that the expressmen may lift it by the handle and set it down lightly, and not be tempted by the light weight of a square box to pitch it pell-mell into the car, and destroy every egg in it. The saw dust should cover the boxes to the depth of an inch, at least; then, if they are not exposed to a freezing temperature, nor to a hot fire, and receive moderately decent treatment, they will go safely thousands of miles. We have sent them safely beyond the Rocky Mountains, to California, to England, and to France. We have packed eggs in such a box when they were first taken from the fish, and keeping it at the same temperature as the water in the troughs have left it, until eggs taken at the same time, and placed in the troughs, were hatching out; and then, opening the box, have found that some of the fish had already appeared, others were just breaking the shells, and all the impregnated eggs were alive, and in good condition. Of course the young fish would not live in the moss, but would die as soon as they appeared. We do not mention this as a new method of hatching eggs, but to show how perfect the means is of sending them. The eggs in the box should be spread as thinly and evenly through the box as possible, taking care that none of them touch the sides of the box, and the moss packed in well (not tightly) to keep them in place. If this is not done the recipient of the eggs will sometimes find them, after a long journey, jolted together into a solid mass, and spoiled. Use clean, bright tin boxes, which are free from iron rust.

CHAPTER V.

Young Trout.

APPEARANCE.—After the eggs have laid in the water from fifty to seventy-five days, according to the temperature, the Trout will begin to make their appearance, the egg appears to be endowed with life, and the motions of the Trout inside " kicking " against the shell to force their way out can be plainly perceived without the assistance of a microscope. At length the Trout forces his way through, head first, or tail first, which ever may happen to be most convenient, and the useless shell floats away down stream. The Trout is then about one-half inch long, and the body proper as thin as a needle; the most prominent features being a pair of eyes, huge in comparison with that of the body, and a sac nearly as large as the egg. This sac is attached to the belly of the fish, and contains food, which the fish gradually absorbs. If the fish are hatched in fifty days the sac lasts about thirty, if in seventy days, about forty-five. At this period of their lives they will work down into the crevices of the gravel and along the sides of the troughs and stay there, nature seeming to give them the instinct, at this weak and defenceless period of their lives, when they are burdened with a load which they can hardly carry, to get out of sight and out of the way of harm as much as possible. At this stage of their growth many curious deformities appear, more interesting perhaps to the physiologist than to the Trout culturist. Some of the fry will have two heads, and some will be united after the manner of the Siamese Twins. A very common deformity is a crook or bend in the Trout, giving it a semi-circular form, so that when it attempts to swim it can only progress in small circles. All those deformed soon die, and may as well be removed from the trough at once, unless you wish to keep them as curiosities. They live as long as the sac supplies them with food; when the sac is exhausted they cannot swim about much to get food, and die of starvation.

The glass which lines the bottom and sides of the trough should lie closely. If there are any openings, numbers of the fry will wedge themselves in and die. As the sac disappears the young fish get larger and stronger. When it is nearly or entirely gone they will begin to rise from the bottom, swim about and forage for food.

NURSERY.—The most critical period in the life of a Trout has now arrived. More, perhaps, die from the time they begin to feed until they are six months old than at any other time. In consequence many different plans for nurseries have been suggested and used. We give that which we have found the most successful. In place of erecting other and wider troughs or boxes for nurseries, the better plan is to put only a few eggs, say five hundred, into each square or nest of the hatching trough. The square is then large enough (with the water raised) to keep the Trout well for a month or two after they commence feeding, and then transfer them into the first pond. This plan economizes space, saves one removal, and the fish do better after a month or two in the ponds than they would in troughs or rearing boxes. It is probably better to remove the gravel from the troughs as soon as the fish commence feeding, because then the troughs can be kept clean more easily, else particles of food will lodge in the gravel, where they cannot be removed. If the water has been well filtered and the flannel screens well looked after, there will not be *sediment* enough in the troughs to require cleaning until

the end of the season. The water must be raised by the
cross-strip before mentioned as soon as the eggs hatch out.
It would be well to fix a small screen in each alternate cross-
strip, which can be done by cutting out a space of eight
inches by two, and nailing a fine screen over the orifice.
This will prevent the Trout from running up and down in
the troughs, and inconveniently crowding together. The
fry are removed from the troughs into the pond by the use
of a small net, such as described among the implements of
the fish raiser. Take them upon this, a few at a time, and
put them in a pan of water; they will swim off the net and
you may draw it from under them. In the pan they may
be carried, a thousand at a time, to the pond in which you
wish to place them. Put them into still water; they will
settle down on the bottom and remain there for some
hours; then they begin to explore their new quarters, and
in a few days will become thoroughly habituated to the
place. The pond should be covered over almost entirely
with boards; this for two reasons: First, to avoid the
fungus growth which will form in summer in the ponds
wherever the sunlight comes. We mean the long, green
fibres of moss which rise from the bottom of the pond, and
which in this locality are called "frog spittle." On a bright
sunny day this becomes detached from the bottom, and
rising to the top of the water floats down with the current
and clogs the screens; besides it fills up the pond so that
the food which is thrown in sinks down into it—the fish
cannot get at it, and decaying on the bottom it fouls the
water of the pond. Second, because the fish require a
shelter in stormy or cloudy weather. In their natural state
we see young fish feeding on the shallows under the bright-
est sun; but in cloudy weather not one is to be seen—they
have sought shelter under stones, weeds, &c. Still the
fish require sunlight, and enough should be given to keep
them in good condition. The board covers are also inci-

dentally useful in keeping down the temperature of the pond. This is apt to rise in hot weather, as much water cannot be let into the pond at first for fear of washing the young and weak fish against the screens by the force of the current. More water must be let in as the fish grow older and stronger.

Food.—Any food of an animal nature which can be minutely and uniformly divided, will serve as food for the young Trout. Liver can be boiled and grated, or raw liver can be chopped up with an old razor very fine and then fed to the young fish drop by drop. The yolk of an egg boiled hard and grated makes a good food ; but in our experience the best food is thick milk or bonny clabber, which can easily be strained through a coarse cloth and thus reduced to atoms. It is not material about removing all the whey, as the water will wash it off, and it does not hurt the fish ; dilute with one-half or two-thirds water. A half tea-cup full of this mixture will feed a hundred thousand fish when they first begin to feed. The best way to feed them is to take a case-knife, dip it in the feed and slirt off the food which adheres in to the troughs ; a very simple way, but one answering all practical purposes. Care should be taken not to feed too much, else the surplus food will remain on the bottom, and decaying there foul the trough. The reason of the difficulty in raising young fish appears to be that they are literally starved to death. The food which we can give them is not natural to them, and it is often given in such coarse pieces that they cannot take it, and sometimes, through the carelessness of a hired hand, they are neglected two or three days at a time. The question will very naturally be asked, " Is it not possible to obtain their natural food?" We answer, yes, in small quantities. The moss and weeds in every stream are full of it, and by placing in the troughs fresh moss and weeds gathered from the stream every day, the young fish may be fed. But this is not practicable. With a large number of fish it makes too much work. It would soon foul the troughs and cause the fish to die, and then, if the young fish are fed on this at first, they will not take the artificial food afterward. Trout, of any age, learn to eat that food which is most abundant around them. Anglers know this by experience, and use the flies which they see on the stream on which they are fishing. It is commonly supposed that a Trout is very fond of grasshoppers, but the Trout in one of our ponds which we have fed for a long time with lights, will not look at grasshoppers, and, strangest of all, will turn up their noses at the fattest and juiciest worms, while the Trout fresh caught out of the stream, which we have put in a pond by themselves to educate, will for weeks refuse the daintiest bits of lights and liver. Hunger will after a time drive them to change their food ; but with the young ones we cannot wait for this, as they will die off before they learn. As the fish grow older and stronger more food must be given to them ; still, when six months old, a bowl full of curd, diluted with water, will answer for a thousand. While the fish are young, feed often ; three or four times a day for the first two or three months, and the oftener the better ; twice a day will do after three months until they are a year old.

A word just here about thick milk or curd. If milk stands a short time in hot weather, or a longer time in cold weather, it sours and becomes thick ; this thick milk is called, we believe, bonny clabber. The process may be hastened by heating the milk, or by the addition of rennet. In that case the product is called curd. The milk naturally turned is best for the young fish, as it is softer and more easily picked to pieces. We have often seen the young Trout, one after another, take and reject small pieces of curd which had been turned by heating, while they would bite to pieces and consume a lump of the naturally turned bonny clabber. When the fish are three or four months old it may be made fine enough by stirring with a spoon, and if there are a few large lumps they will not go to waste, as the fish will pull them to pieces. Curd is best for the larger fish, as it is more compact, and holds together in lumps. The bonny clabber may be given to the fish until they are a year old, but after that it is generally more economical to feed them upon liver or fish. A change of food is good for fish, as well as for all other animals. But as this whole question is yet unsettled, or more properly the art yet in its infancy, we give only our practice.

A feeding platform in the ponds may be used with advantage. This is simply a platform of boards, two feet by four, placed in the middle of the pond, and raised a few inches above the bottom ; it will also serve incidentally as a cover for the young fish. If you throw the food over this platform, all, if not taken before it reaches the bottom, will fall upon the platform, and as this can more easily be cleaned than the bottom of the pond, there is less liability of fouling the water ; the fish will also take food better from a clean bottom than when the food lodges in the mud or weeds.

There will be a great difference in the growth of the fish noticeable after the first few weeks of their existence. Some, of course, will be larger and more vigorous than others from their birth ; but of those apparently of the same size and health when one month old, some at six months will be four times the size of others ; this, too, when grown in the same pond and under the same circumstances. They will begin to eat each other when very young. A Trout only a few weeks old begins to show symptoms of fight, and will kill his weaker brethren when they get in his way by biting a piece out of their tails. In two or three months, when some of them get to be double the size of others, they will swallow each other whole. We have taken a Trout one inch long out of another only two inches long. It would seem to be advantageous, therefore, to sort them out every little while, and put the same size by themselves; but in practice this is very difficult, and the less a Trout of any size is handled, the better; besides, if they are fed well, they lose their disposition to eat each other. Therefore, the Trout of each year may be left by themselves with very little probability of losing more by cannibalism than would be killed in sorting out and removing.

The covers should be kept on the first pond at night, and in dark, cloudy weather until September, at least, when the Trout will be from two to five inches long and able to take care of themselves. Even then it is as well to have the covers at hand and put them on in case of a rain storm, since we often find after a storm numbers of young fish dead in an exposed pond; therefore, we conclude, although we do not know the philosophy of the thing, that they need a cover in rainy weather. After September there is no difficulty in raising the fish—they all appear hardy and strong; perhaps it is because all the weak ones have died off, and none are left but those able to " stand the pressure."

HARD TO KEEP.—There will always be a difficulty in so arranging ponds, screens, outlets and inlets as to keep the young fry in their proper pond. The water is very apt to work holes around the screens, or rather around the boxes containing the screens.. The young fry will work their way through a wonderfully small hole, no matter how long the distance may be. They will also get through between the screen and the socket, unless they are very well fitted together, and worse than all, wherever there is a crack into which they can get their large heads, they will

put them in so tightly that they cannot extricate themselves, but will die. In short, wherever you can run the
big blade of your jack-knife, there the young Trout will go.
In making a pond for them, it is best to beat the edges
with a spade until they are perfectly smooth, or, better yet,
to put a board around the edges to the depth of a foot.

CLEANING SCREENS.—If the screens are not kept well
cleaned, two ends follow : First, the water runs over the
top of the screens instead of through them, and the young
Trout will escape; and second, when the screens are
taken out to be cleaned a rush of water follows their removal, carrying away with it numbers of Trout into the
next pond. Whenever you are going to clean the screens,
drive all the Trout from their vicinity, then take the screens
out and wash them with a stiff brush. They can be more
easily cleaned by turning the dirty sides downward and
knocking them against a stone. Try it once and see how
easy it is, and then—don't do it again, if you wish to save
your screen. It will be also seen now why it is necessary
to have an additional supply running into the second and
third pond. A supply of water sufficient for them would
be too much for the first.

DISEASES.—This part of fish-raising is least understood
as yet. After the egg sac is absorbed and the fry begin to
swim about, a sick one is very easily distinguished. The
healthy Trout swim in the current with their heads up
stream, darting about here and there after minute particles
of food. The diseased ones wander about listlessly, swimming round and round continually. They may also be
known by the size of their heads, which appear much
larger than their bodies. The head of a young Trout is the
largest portion of the fish, even when well, but when sick
it appears to be all head. When they are thus affected we

4

don't know what is the matter with them, and how to cure them. Before the food sac is gone the Trout is often afflicted with a swelling over the sac ; a membrane forms there, swells out large and is filled with a watery substance. We call the disease the "dropsy," for want of a better name. Sometimes the Trout may be saved by making an incision in the swelling and letting out the water ; but as with care only a few of them are affected in this way, it is better for the fish culturist to hatch more eggs than he expects to raise than to bother with a surgery he does not understand. In other words, hatch more than you want, and keep the strongest and best.

In review of this chapter the main facts are that the infant Trout are hard to raise. It is almost as if a great rough man should attempt to raise a thousand infants' deprived of their parents at birth. Most of the infant Trout die of absolute starvation. They do not get food enough. That which is given to them is not cut fine enough, or for some cause they will not take it.

CHAPTER VI.

ADULT TROUT.

SUPPLY OF WATER FOR GIVEN NUMBER OF TROUT.—
This has never yet been accurately determined, and we do
not know that any general rule can be given applicable to
all times and places : because the supply of water required
for any given number depends very much upon the temperature of the water,—a given supply in cold weather
sustaining many more in good condition than the same
supply in hot weather. It is the same with Trout as with
mankind. If many people are packed together in a close
room, they will soon begin to suffer ; but will not feel the
bad effects so soon in cold weather as in warm. Now the
water contains the air upon which the Trout lives, and the
supply for any given number depends also upon the
amount of air which is in the water. A still and smooth
flowing stream, with little vegetation in it, contains the
least amount of air. Hence the value of a fall of water
between the ponds if the supply is small. The volume
of water required depends also upon the shape of the
ponds and upon the size of the fish. We can only say
about what quantity is necessary and leave each owner of
ponds to observe for himself whether more or less fish do
well in his locality, and under the circumstances, of ponds,
supply, &c., which he has made. It must always be borne
in mind that the larger the supply of water the better for
the Trout ; and the Trout-breeder on a large scale will find
better success with small ponds and large supply than in
any other way. For ten thousand fish the water supply

should not be less than seven inches square (that is, forty-nine square inches) and would be still better if it was seventy-five square inches. A less supply will perhaps do; but with it there is danger of disease and death to the fish. We will say then a supply of water filling a pipe five inches square (making twenty-five square inches) for the size of ponds shown in plate on page (14) calculated to sustain five thousand fish in the second, and two thousand fish in the third ponds. The first pond to receive six or eight thousand young fish, need not have more than two or three square inches of the water. This estimate of number of fish is purposely made low. More fish *may* be able to live in your water with the supply mentioned; but the number given *certainly* can.

GROWTH OF TROUT.—It is impossible to tell the age of a Trout by its size, as its size depends very much upon the quantity of food which it obtains. It is a general rule that with good feeding a Trout three years old will weigh one pound. They have been known to live for years at the bottom of a well, where the supply of food must have been extremely limited, and remain through all those years, apparently at the same size. Then again, with good feeding, they will more than double their weight in a single season. Trout will not grow so fast in swift running water as in a pond. The largest Trout are never caught in narrow parts of the stream where the water runs fast. But where the rivulet swells out into a dark and still pool, there the patriarchs are found. We presume that the largest Trout now taken in this country are found in the lakes of Maine. Some will grow much faster than others under any circumstances. A few will always look lean and hungry no matter how much they are fed, and others seem to have a peculiar knack of getting fat. Still the rule of good feeding applies equally to all. They will not grow so fast when three or four years old as before ; that is, the rate of increase diminishes with age. The average age of a Trout is perhaps twelve or fourteen years. On this point we cannot speak with certainty. When I first commenced building ponds, five years ago, a large Trout was caught from the brook, which I supposed then to be three or four years old. From its size and beauty it was deemed worthy of being distinguished by a name. Helen, as she was called, throve well in the pond to which she was transferred, and increased in size and comeliness : at the beginning of the present year she seemed to be as well as ever and as large; weighing nearly four pounds. But during the year, without any perceptible sickness, she has diminished in size, her back has turned gray and grizzly, and she presents a stunted appearance, which says as plainly as words to the spectator, this Trout is old and gray and wrinkled. Judging from this and from others in our possession, we suppose a Trout to be in its prime when it is from three to ten years old. The size is altogether a question of food. On Long Island where they have access to the salt water and feed on the numberless small fish and crustacea abounding in the sea, the Trout are notoriously large, while in the mountain stream, where the food is scarce and precarious, it is just as well known that the Trout are small. The size to which a Trout may grow is not very well settled ; so many " fish stories " have been told that discredit is thrown even upon well authenticated assertions. Trout may in exceptional cases and in large waters attain the weight of eight or ten pounds, but the largest one I ever *saw* weighed five pounds and two ounces ; and a four pound Trout is generally considered to be of pretty good size. This question of size is interesting rather to the sportsman than to the Trout farmer. It is universally considered by all old Trout-eaters that small Trout are the best—say from one-quarter to one-half a pound. A better

market may always be found for fish of this size than for
any other. There is only one market in the United States
where there is a demand for very large Trout, and that is
New York, where the largest Trout sell the most readily.
Besides, fish of this size (small) are the handiest to manage
in the spawning bed, and more of them can be raised. If
the spawn is extracted by hand, the difficulty in handling
a two pound Trout is very great and increases very fast as
the fish grows larger. Not only is it a great trouble to
handle the large ones, but the danger of killing them is
much greater; so that, in our opinion, from one-quarter to
one-pound weight is as large as the fish farmer should attempt to grow his Trout, unless from motives of curiosity
to see how large they will get to be. As to the flavor of
a Trout, nothing need be said in its praise, as its excellence
has long been conceded; but much has been said about
the inferiority of flavor in Trout raised artificially. There
is no foundation for such an opinion. The angler, of
course, thinks no Trout equal in flavor to that which he
catches himself. After he has been tramping all day in the
open air and gathering a raging appetite from his exercise
and sport, he will cook a Trout over a smoking wood fire,
eat it half raw and half burnt, and declare it the most de-
licious morsel in the world. It is not very reasonable, at any
rate, to suppose that a lantern-jawed, slab-sided, long-bodied,
half-starved Trout out of a mountain stream is equal in flavor
to a well-fed, fat and sleek Trout from the artificial pond.
As to actual experience I cannot tell any difference in flavor
myself, neither did I ever meet any one who made the trial
under fair circumstances who could detect the difference.
This statement must not be misunderstood. There is a
great difference in the flavor of Trout taken from different
parts of the same stream, and at different times of the
year, even from opposite sides of the stream. Still the
same trouty flavor is distinguishable in all, and the flavor

is the same in the Trout taken from the artificial pond as in that taken from the stream—varying no more and no less.

A word may be said here about the distinctive marks of Trout. We consider that the differences which are found are caused by food, climate and water. The fringe marks or bars found on the young Trout soon disappear. In adult Trout the markings vary considerably. Some will be more highly-colored on the belly than others, and on some the spots will be brighter. Some look dull and dark ; others light-colored and brilliant. I have myself caught the silver or Schoodic Trout, as it is called, in the Caledonia creek. This Trout is commonly supposed to be confined to one locality—the Schoodic Lake—and is also called the land locked Salmon. We have still in our ponds a number of the same species, caught from our creek. Trout can be bred to any color by feeding and the use of proper ponds, and we believe that in the future they will be bred to color, shape, flavor, etc., with as much nicety and certainty as a cattle-fancier breeds his animals.

The color of a Trout is sometimes a matter of optical deception. Old fishermen affirm, we think truly, that a Trout is always the color of the bottom over which it lies ; and that in passing from one color of bottom to another, it will change its color in a minute. The Trout in deep and shaded pools are notoriously deeper in color, or rather darker than those in shallow, bright waters ; and they not only look darker while they are in the water, but stay darker when they are removed. The Trout-raiser must make his ponds accordingly : shallow and exposed if he wishes light-colored Trout ; deep and shaded if he wishes a darker color.

FOOD.—The question of food for Trout has not yet been entirely solved. By this we mean that it is not yet settled

what is the best food which can be obtained cheapest and
in the greatest quantities. This question is important because the profit of Trout-raising depends upon it. All other
circumstances being equal, he who can obtain the cheapest
food will make Trout-raising pay. In France and Ger-
many dead animals are gathered from the farms around
the fish establishments and made into pates, or pies, which
are fed to the fish as wanted. However good this may be
for the fish it is somewhat repugnant to the taste of the
fish eater. In this country we pursue a cleaner method.
The pluck of animals killed (that is the lights and liver) is
obtained from the butchers. This food can be obtained
fresh at least once or twice a week in most localities and
kept fresh by means of an ice house. In fact Trout will
not eat decayed or spoiled meat unless they are very hungry. They are very dainty in their tastes and will often
go hungry rather than take anything which they do not
fancy. We feed meat to them raw. They have never
been used to cooked food in their natural state and we do
not see that it is any better for them. The lights should
be given to the larger fish as it cannot be chopped so fine
as the liver and is more apt to hang in strips or strings.
The liver which can easily be cut into small pieces may be
fed to the smaller fish. A Trout will sometimes choke to
death ; they are so greedy that they attempt to swallow a
very large piece of food. Sometimes this sticks in their
throats and kills them. Often it is caught in their teeth
and thus prevented from going down the throat, or it gets
into their gills and stops their breathing. They will, when
choking, come to the top of the water, and may sometimes
be saved by taking the piece out of their throats, or push-
ing it down. But the best remedy is to chop the meat fine,
say one-half or one-quarter inch squares for two and three
years old. No machine which we have ever tried would
do the work of chopping to our satisfaction. A sausage

machine runs the food together and mashes it, and the meat-cutters, which do the best, require cleaning and sharpening so often that they are only a nuisance. The best thing we have found is a butcher's block, or log of wood two and a half feet high on which to cut, and a very heavy knife or light butcher's cleaver. These instruments are very simple, not liable to get out of order, and do the work required of them in the best manner, and with no more labor than a machine would require. Sometimes two or three knives are fastened together to make the work go more expeditiously; but one is best.

Any kind of meat is good for food. Trout are carnivorous and will not eat vegetables of any kind that we have ever tried. We feed them lights and liver because it is the least expensive food we can find in large quantities, and answers a very good purpose. In their natural state Trout feed upon insects of all descriptions which abound in or near the water; worms of all kinds, from the angle worm to the caterpillar, which the wind shakes from the trees bordering the stream into the water, are eagerly taken. Flies of every kind which either drop down upon the surface of the water to lay their eggs, or may happen to fall into it, are quickly devoured. Young fish of any kind which may be in the stream serve for food; so do the grasshoppers and beetles which fall into the water, and even the crawfish is not spared. If any one will examine the bottom of a good Trout stream carefully, he will find every stick, stone and bunch of moss in it covered and filled with insects of various kinds. If you look at the bottom of the creek, also, when it is free from moss and sticks, you will see that in the summer time it presents a curious mottled appearance, as if it were having an eruption of some kind; these protuberances are caused by the eggs of water flies, which, after a time, rise to the surface, and then breaking their shell or case, for the first time,

spread their wings and fly away. On this larvæ, before it
it has assumed the fly-state, the Trout feed ; and these eggs
of water flies, together with minute insects and worms are
the special food of the very young Trout.

Fish of any kind are a very good food for Trout. If they
are small they may be put into the water whole, the Trout
will take them all the better if they are alive. Any coarse
fish which can be obtained cheaply and in sufficient quantities may be chopped up fine and used as food. Curd also
may be fed to adult Trout as well as to the young, and
they will thrive upon it. Sometimes maggots are used for
food. A maggot factory is very simple, but also very
nasty. Meat can be kept in an open box until it is nothing
but a mass of maggots, or a piece of meat can be hung over
the pond and as the maggots form they will drop into the
water. The maggots certainly do not contain any more
nutriment than the meat, and the bad smell attending the
operations is argument enough against it. It is just as
easy to have everything clean and nice, and the fish certainly will be no worse for it. As we said before, they
will not eat carrion unless pressed by hunger. They will
eat a live Trout, but we have never known an instance of
their eating, or even touching a dead one. If any way
could be devised of raising flies, or shrimp, or various kinds
of insects (their natural food) in sufficient quantities and at
little expense, this would be the best of all. A change of
food would also do them good, but we find that they will not
readily change their food. This bears upon the question whether Trout will take flies out of season. It has long been a
matter of dispute among anglers, whether Trout would
take only particular flies at particular seasons of the year,
or whether they would take any good-looking fly at any
time. My experience is that when one kind of fly is thick
upon the water the Trout will take no other. When there
are very few flies and food is scarce they will take almost

anything in the shape of a fly which presents itself. On Caledonia creek an angler with only a few general flies would make a very poor catch. The Trout there depend chiefly on flies for food and learn to be discriminating. On the Long Island streams where they feed chiefly on salt water fish and to take flies only as occasional delicacies, they may very easily be deluded.

As to the quantity of food necessary for a given number of Trout. This is difficult to give exactly as it will vary with the size of the fish and the season of the year, more being required in moderate weather than when it is very hot or very cold. For one thousand three year old, about five pounds of light or liver per day; for two year old three pounds; but a very little trial will show just how much to feed them. Feeding once each day will keep the Trout over one year old in good condition. Feed slowly, and as soon as they begin to refuse the food stop feeding them, then you have the measure and feed a little less than this quantity every day. We say a little less because we have known cases in which owners of ponds being over anxious to fatten their Trout, have killed them by over-feeding. Still this does not often happen, especially if they are fed regularly. A Trout after long abstinence will gorge himself to repletion; but will not kill himself to-day if he is reasonably sure of to-morrow's dinner. All animals appear to be wiser than men, and it is very seldom that they will eat enough to do them injury no matter how much may be given them.

The Trout will get so tame after a time that they will take the food out of your fingers, in fact they will take the fingers too. Their teeth are sharp and make scratches like needles. They may be taught to jump for their food by holding it a short distance above the water, or may be made to come up and take it out of the pan you are holding. Feed in the middle of the day when the sun is well up, any time from ten to three is good.   Make it a general rule to feed slowly and give them as much as they will eat without wasting.

The question has often been asked us whether salt food agrees with them.   We only know that our fish will not take it much, and it does not seem to agree with *them*. Still as Trout will live and thrive surprisingly in salt water we suppose they will, in time, relish it, although eating food found in salt water and eating what is commonly termed salt food are things very different in degree.   We do not recommend its use.   For many, curd is a much cheaper food for adult Trout.   The farmer who keeps cows will find it an advantage to turn his cream into butter and feed the curd to his fish.

TEMPERATURE OF WATER.—The colder the water is, down to forty degrees, the better will the Trout do.   They will die in the ponds if the water raises to sixty-eight degrees, unless there is a spring in the pond, or colder water into which they can get.   We have often heard or seen the statement that fish could be kept in a frozen state a long while, and then thawed out and be as lively as ever.   Our experience says no.   Fish may be frozen, so that a thin scale of ice forms over them, and so long as they can be bent they will thaw out and will recover ; but if they are once frozen solid or stiff throughout, they are dead, and cannot be brought back to life.   If the ponds freeze over in winter, it is no sign that the water under the ice is below thirty-two degrees.   If it was, the water in the ponds would freeze solid.   Unless the water is taken close to a spring and much water runs through the ponds, the surface *will* freeze over ; but this will not injure the fish, as the water below will be much warmer than the temperature of the atmosphere ; and the ice which forms over the pond serves to keep the water below from being made colder by contact with the air. If the water is so sluggish as to be likely to rise above sixty-eight degrees in summer, the ponds may be shaded in some way. Trees and bushes look very nicely about the borders of the ponds, and are valuable so far as ornament is concerned. But there are certain objections to their use which will banish them pretty thoroughly from the grounds of the practical Trout-raiser. One objection is that the leaves, in autumn especially, clog up the screens, and demand constant attention to prevent an overflow of the water and Trout. Or the leaves fall to the bottom of the pond, and decaying there, foul the pond. The roots of the trees also will force their way towards the water, and break the walls or banks of the pond. If it is necessary to shade the ponds, floats may be used, made of boards nailed together and moored in some convenient place ; but the best plan of shading is by light covers placed on beams running across the pond. If the ponds are very large, the floats will have to be used. But the ponds must not be made large. We have said that Trout would not live in water which was raised above the temperature of sixty-eight degrees, and would do better in water at forty degrees. This settles the question as to how far south the Trout will live. It will live as far south as the temperature of the water in summer will allow. As to the exact latitude that depends upon elevation, &c. ; but there are few Trout, if any, to be found in the Southern States.

DISEASES AND ENEMIES.—The diseases to which adult Trout are subject are very little understood. At least the cause of the diseases and their remedies. Sometimes a Trout will be observed to have a white fungus growing upon it in spots. This will spread over the fish until it dies. Sometimes the fish will turn to a black color. This always seems to be an indication of blindness, as we

have never observed this peculiar color unless the fish was
partially or totally blind.   The fungus which grows upon
the fish is probably not a disease, but is caused by, or is the
indication of a disease. Nothing is known about remedies.
If only a few Trout are affected, take them out, as they
will be sure to die; those covered with spots very soon,
and those blind, in a few months, of starvation.   If the
Trout begin to die in numbers, change them to another
pond, if possible, or give them more water.   This is all we
can do for them.   The dead Trout should be taken out of
the pond as fast as they are discovered.   They will rise to
the surface only in very rare cases, but generally sink to
the bottom, and if there is much moss in the pond they are
lost to sight, and decaying on the bottom, foul the pond.
If there is much sickness among the Trout, we generally
consider it a sign of insufficient water.

There are but few enemies of Trout in artificial ponds.
If the ponds are near the house, and people constantly
about them, there will be no trouble with the birds which
usually prey upon fish—such as the kingfisher, fish-hawk
and crane.   Even if the ponds are some distance from the
house, the water will probably be too deep for the fish-hawk
and kingfisher to do much mischief, as it is only in shallow
water that they can be certain of their prey.   Cranes will
wade into the water and take all that, comes within
reach of their long bills—whether frogs, snakes or fish.
But they are very few in number, and the Trout are wary.
If any of these birds appear, shoot them; and as there
are very few of them, at any rate, you will not be troubled
much.   Muskrats sometimes get into the ponds.   They
are vegetable eaters and do not eat the fish.   Still, it is
bad to have them around, as they make holes in the banks
of the ponds and let the water off.   A few traps will
soon dispose of them.   It may be worth while to mention here the manner of catching them.   Find out the

places where the muskrats go into the ponds. They will make a little bare path, or run on the edge of the bank, by always going in and out at the same place. Then set a trap (a common game trap, such as is sold in all country stores) in the water, so that the plate of the trap will come in the middle of the run and about a half inch under water, taking care to place the jaws of the trap in such a direction that when shut they will be in a line with the run. Then stake the chain into deep water. No bait is necessary. If any bait is used, a sweet apple or parsnip is good. The muskrat comes through his run, steps on the pan of the trap and springs it. He tries to take it with him to the shore. If he succeeds in doing this, he will likely get out in some way; for instance, if he is caught by the leg, he will sit down and knaw it off, in order to get free. But as the trap is staked out into the water, he cannot get to shore, and will be drowned by his struggles and by the weight of the trap, for he cannot survive under water very long without rising to the surface for a supply of air.

Water snakes cannot do any damage to the large Trout, and even the small Trout are too much for them, unless they are very thick, or are in a very small pond. We have seen the large Trout in our ponds dart and apparently attempt to bite a snake, which was passing through. They exhibited no fear of the snake; but the snake certainly did appear to be afraid of them. Undoubtedly they will eat the small fish if they can catch them. We have often opened snakes, caught about our ponds and creek, but never found any Trout in them; so the danger from snakes can not be very great, except to the very young fish. Frogs have a very bad name; but we think, do not eat the fish very much, although they will certainly eat all the little ones they can get hold of. Even if they do no injury, they are not of any advantage, and may as well be disposed of.

Crabs, or cray-fish, as they are more properly called, very seldom eat the young fish. They will lie on the bottom of the pond, hidden in the mud, with the joint of the claw wide open and ready; then if any unfortunate Troutling passes within reach, his doom is sealed. Cray-fish do much more mischief by their burrowing propensities. They will make holes out of the pond, or from one pond to another, through which the water escapes, and very often the young fish also. The cray-fish is the scavenger of the water, and it may be a question whether a few of them will not do as much good, by disposing of decaying animal matter, as they do harm, by destroying a few fish. The greatest fear of all fish-raisers is that their fish will be stolen at night. Well, there is the same danger here that there is in any other property, and no more. In fact, not so much. The risk of the States Prison is too much for any one to incur for the sake of a few fish : and then there is only one way in which the fish can be obtained. They cannot be taken out of the ponds with a scoop (or scap) net. If any one will try this a few times, even in broad daylight, he will soon be convinced of the fact. A few old logs, stones and branches of trees, strewn on the bottom of the pond, will make it impossible to drag the pond with a seine. Catching them by hook and line is the only means; and if the fish are well fed daily, it will take more time to catch a mess than thieves can usually spare. Trout also find enemies in their own kind. The only way to stop them from feeding on each other is to give them plenty of other food. It may be as well, perhaps, not to feed them on small fish, unless they are chopped up fine, for this reason : Trout soon accustom themselves to certain kinds of food, and will refuse anything strange. If they get into the habit of feeding on small fish, they will not be likely to make a distinction between Trout and any other fish. Certain old Trout also

become unusually destructive to their brethren.   Like the
" rogue-elephants," and the " man-eaters," among the lions,
they become morose and sullen, live apart from the rest,
and make war upon everything around.   When you find
one of this kind, spear him at once, as there is no cure,
and he will invariably destroy more than he is worth.   It
may be worth while to mention here how one Trout eats
another : An old Trout will catch another, in some cases
one-half of its own size, by the middle, and with its strong
jaws hold it fast and swim around with it, while the pri-
soner worries and struggles to get free.   This performance
lasts until the victim gets loose or is exhausted.   I have
seen one carry another round in its mouth, for half a day.
If the little fellow gets free, it is usually only to die a lin-
gering death ; for the breaking of the skin is fatal.   When
it is exhausted, the old rogue, dropping his victim, which
until this time he has held by the middle, seizes it again
by the head, and slowly swallows it whole ; the operation
sometimes taking several hours, and while in progress
making the fish look as if it had no head, but only a tail
at each end.

In some localities mink are very destructive.   These
animals are particularly to be dreaded because they do
not only take what fish they want to eat and then leave,
but will take out fifty or one hundred before they stop, and
having found a well stocked pond, they will resort to it
again and again.   The best way to trap them is as follows :
Make a box eighteen inches long by six inches broad and
deep, leaving one end open, set a common game trap (such
as used for catching muskrats) in the open end of the box
in such a position that when the jaws are closed they will
be in a line with the length of the trap.   If it is set crossways it will be apt to throw the mink out instead of catching it.   Put the bait in the further end of the box—a piece
of meat or a dead fish will answer for bait—set the trap and

4

cover it over with a large leaf. Now, there is only one way for the mink to get at the bait, which is by walking over the trap. Some Trout-breeders also try to raise mink for profit as their skins are valuable; but their habits of eating fish and their custom of getting out of almost any box or yard in which they are confined do not make them agreeable neighbors for the Trout.

The fish farmer can always tell by looking at his Trout in the morning whether they have been disturbed during the night. If they have been molested, whether by birds, mink or men, they will appear excited and frightened. The water will be discolored by the mud which they stir up as they dart back and forth near the bottom, and the Trout will be nearly all hidden under stones, in the moss, etc.

There is one kind of Trout which we do not possess in ponds, of which we would very much like a specimen. We mean the Trout which comes to dinner at the sound of a bell, or at the call or whistle of his feeder. Many writers about fish tell us to avoid all noises around the ponds lest they frighten the fish, and to be particularly careful never to fire a gun on the grounds lest the delicate ear of the Trout should be too much affected. Now, having made somewhat careful experiments with the view of writing this, we would only remark that Trout cannot hear the sound of a bell, nor the voice of their feeder, nor even his whistle, neither will they stir one fraction of an inch at the sound of a gun fired one foot above their heads; but the sight of a Trout is very keen, his eyes are magnifying glasses, and make everything look larger, and at the same time indistinct. His enemies are to be avoided by the aid of his eyes, and the Trout starts and runs at every sudden motion, whether it is the shadow of the angler, or the falling of a leaf upon the water. But this is not exactly Trout breeding; let us return to the subject.

TRANSFER OF FISH.—In the spring of the year when the new hatching is ready, the Trout must be moved, those from pond No. 1 to pond No. 2, and from No. 2 to No. 3. The manner usually adopted is to take out the screens between the ponds and drive the fish down. Care must be taken that all the fish are removed from the pond before the next size is let in, as one large one will create havoc among a batch of small ones. This method reads very easy in print, but there will be found some difficulties in practice. It is hard to get all the Trout out, unless the water is drawn very low and the fish then taken out with hand-nets. They will hide under a stick, stone, or clump of moss, and you may think that every fish is out of the pond when there are, in reality, dozens of them left.

OTHER FISH WITH TROUT.—It is not well to have other fish in the same pond with Trout, they will probably destroy one another. Fish of any sort will eat the young of all kinds. Even the harmless and innocent looking goldfish will take young Trout with a relish. A few sticklebacks will probably get into the ponds, but they will do no hurt unless they get among the babies two months old ; the large Trout will soon clear them out of their vicinity. Let the pollywogs wiggle their way in peace, and when they get to be frogs sell them or eat them.

CHAPTER VII.

Taking Eggs.

Season for spawning.—All fish of the Salmon kind spawn in autumn or winter. Trout commence to spawn about October. The colder the climate is the earlier they will spawn. In our stream (Caledonia Creek) the Trout commence to spawn about the 12th of October; the water standing then at about forty-eight degrees. In our ponds where the temperature, at that time, is a few degrees higher, they begin to spawn about the 1st of November, and they cease spawning about the 1st of March. The length of the spawning season depends upon the equality of the temperature of the water. In streams where the temperature does not vary much, the length of the season is three or four months, sometimes more, and in cold mountain streams it only lasts two months.

Signs of Spawning.—As the season of spawning approaches, the difference of sexes shows more clearly. It is very hard in the summer to tell the difference between a male and female Trout. By handling them much and watching them closely the Trout-breeder comes to know the male and female apart almost instinctively; but he would be puzzled to tell just *how* he knows it. The male is generally sharper jawed than the female at any season of the year, and lines drawn from his shoulders to his tail would be straight without any bulge in the middle, while the female has a rounder jaw, and even in summer is slightly protuberant in the middle. These are general signs, and by no means universal. It is only in the spawning season that difference of the sexes can be told with any certainty. As this season approaches the differences become more marked. The difference in size may be more easily distinguished, as the eggs grow large and fill the belly of the female. It will not do to mistake food for eggs. A Trout recently gorged with food looks just like a female full of eggs ; but the food soon disappears as, a Trout is an animal of quick digestion, while the swelling caused by the maturing eggs gets larger as the spawning season approaches. The colors of the fish, also, are at that time a guide. The female turns to a dark and sombre hue, while the colors of the males grow very brilliant.

NATURAL SPAWNING.—As the spawning season approaches, the Trout seek places in the creek adapted to the purpose. The places sought after are a pebbly bottom in shallow water close to the spring or head waters of the creek. Trout will work their way up over the shallows of a stream clear to the inlet ; but if there are springs in the bottom of the creek (which is the case with almost all creeks) they will invariably spawn there, without going up to the inlet, or if they find a shallow place with gentle current and gravel bottom anywhere in the creek, they will spawn there. (Very few of the eggs laid in such a place will come to maturity unless there happens to be a spring there). The males sometimes go up the stream first, not always. At this season the males engage in fierce contest for the possession of the female. These battles often end in death to one or both of the combatants. That these battles are fierce, the deep wounds left on the dead bodies of the slain will bear witness. I have known them to fight for two days, and then both be killed. However, when they are once mated the battles cease and the pair are hardly ever seriously interfered with. Intruders in any quantity come

around, seemingly out of curiosity or fun ; but, no matter
what their size, they leave as soon as the husband, for the
time being, darts at them. These intruders are, perhaps,
waiting for a chance to devour some of the stray eggs
which the female drops. The male and female being
paired, go to the chosen place. They lie side by side together when not disturbed ; but the male is occupied most
of the time in driving off intruders who are continually
swimming around. It is very curious to see a little male
with a big female in charge. Usually the little Trout
clear the way for the large ones without a show of resis-
tance. In the ponds when the Trout are fed the largest
ones get the meat while the little ones get out of the way,
and swim to the further side of the pond, and even if the
meat is thrown where they are they will not take it until
they have waited to see whether it is not the pleasure of
the big fellows to come and get it. At the spawning season all this is changed, they will attack a Trout thee times
their size if he comes within a few feet of the female.
Very often while the male is driving off one, another one
on the opposite side will make tender advances to the female ; quick as a dart the proper husband returns to chase
the gay deceiver. In fact his time is fully occupied with
chasing off intruders. Very often if they are too numerous the female will dart from the nest over which she
hovers, to help her chosen mate. It may be imagined
that there is not much time for love-making, however, one
compensation is that there are no longer any battles. All
intruders literally " turn tail " as soon as chased. They
seem to recognize the rights of the married pair, but act as
if they could not restrain their propensity for harmless
mischief. Of course it only looks so ; the intruders are
after eggs. We do not know of any animals which enjoy
a sense of the ludicrous, or what is commonly called fun.
Still there are few sights more interesting to the lover of

nature than the spawning of Trout. A nest is made in
the gravel by the female. It is simply a shallow hole
about six or eight inches in diameter and about two or
three inches deep. This is made by the female diving
down at intervals against the gravel and as she comes up
giving it a *slirt* to one side with her tail. Nearly the same
motion as may be often observed when the Trout dive
down on the bottom and rub their sides against it to free
themselves from parasites. This dipping motion is continued for some days until the nest is large enough to suit
her. After lying over this some time the female is ready
to emit a portion of her eggs. The male lies by her side
while she does this. However busy he may have been
in driving off interlopers, he seems to know by instinct
when the female is ready to emit her eggs and is always by
her side. At the same time she emits her eggs he emits
his milt over them. They do this with a curious curl upward, which every Trout-breeder should see for himself.
I do not know that I can describe it so as to make it understood ; but as they emit the eggs and milt lying side
by side, they start forward and upward. Very often the male
and female lock jaws together and their heads slowly rise, apparently trembling with excitement and emitting eggs and
milt until a nearly vertical position is gained, still lying
over the hole, then, the eggs and milt being all emitted,
they fall away from one another and the male retires to
some secluded spot where he remains five or ten minutes
resting. This interval the female employs in covering her
eggs. She will *slirt* in with her tail all the stones of proper size to be found near her nest, and if there are not
enough to cover it to her liking she goes above, and, pick-
ing out a particular stone, works it down backward between the two ventral fins. This labor she continues until the eggs are completely covered. After five or ten
minutes the male pays her a visit to see how she is getting

along. He looks around a little, eats a few of the eggs if
he can find any uncovered, and then retires to his work
again, where he remains two or three hours with only occasional visits to the female before he recovers from the
exhaustion which he has undergone. The female does
not seem to rest, she continues covering the eggs and does
not then leave the place. The reason for this is that she
has not yet emitted all her eggs, for Trout occupy some
time in their spawning, laying their eggs at intervals, as
they become ripe. Observers differ as to the length of
time occupied in spawning. My own opinion is that the
time is not usually more than three days, although sometimes extending to six days, the female covering the eggs
as she emits them. When it is understood that some of
the eggs do not sink into the nest, but are carried off by
the current, and that only a few of every batch escape the
jaws of their parents, and of the many Trout swimming
around the spawning place, one may begin to perceive the
advantage of artificial methods. To make the matter still
worse ; after the nest is finished, the parents gone, and the
eggs nicely hatching, another pair come along intent on
similar business. The female sees the place where the first
has laid her eggs, and, fancying it a good place for her
own nest, begins to make one there. As soon as the eggs
are uncovered, leaving all other business, the pair eat up
all they can find, and then proceed to lay their own eggs,
only, perhaps, to be served in the same way by another
pair. When it is considered, also, that all kinds of water-
fowl are fond of these eggs and diligently search after them,
and that in the spring time the young fry furnish a large
proportion of food for the older ones, the wonder seems to
be, not that there are so few Trout in our streams, but that
there are any left. Another cause of the rapid diminution
of Trout in settled countries, is the tame ducks which are
allowed on the stream. They wander at will peacefully

up and down the stream, explore every foot of the bottom, turning over the gravel with their long beaks, and leaving very few of the eggs to hatch. The number of spawn which a Trout will give has been variously estimated. They will commence spawning at two years old if well fed and large. It has been asserted that eggs have been taken from a Trout one year old, or rather taken in the winter of the same year in which it was hatched. This may be so, but it is more interesting in a physiological point of view than for any practical purpose, as there are so few that it is not worth while to take them. A Trout two years old will give from two hundred to five hundred eggs, a three year old from five hundred to one thousand eggs, a four or five year old from one thousand to two thousand eggs. This is only an approximation, as the number of spawn depends upon the weight and health of the fish, and not on its age. In some cases the number of eggs is much greater, but four thousand is the most I have ever taken from a Trout. In estimating the number of spawn from a given number of fish in a pond, it must be remembered that some are barren, and some diseased, and some, perhaps, will not go up into the race. So that the average yield of two and three year olds, (females only counted), will not be over five hundred, of four and five year olds, not over one thousand each. The proportion of males to females in a pond should be about one half. Not so many are necessary to fecundate the eggs, and it would be an advantage in one way to have fewer, since then there would not be so much fighting in choosing partners, and as all the females do not spawn at once, one male would be enough to serve several females; but, on the other hand, the males seem to run out of milt before the females get through laying their eggs, and towards the close of the season it is often difficult to obtain males with milt enough to fecundate the eggs; so that it seems better to have in the pond an equal number of males and females, thereby giving more chance of saving some of the milt till the last of the season. The males are very amorous and will pair again and again. It very often happens that some of them die from the exhausting effects of the season. The best we can do is to have an equal number of males and females, and take the chances.

TAKING SPAWN BY HAND.—There are two methods in practice for taking spawn. The old method, which we will explain first, is that most generally used hitherto, and with it a good degree of success has been attained. The Trout will not spawn in the ponds where the bottom consists of large stones or weeds; but if there is any sand or gravel anywhere on the bottom of the ponds they will spawn on it. Therefore be careful to have only the raceway, where the water enters, covered with gravel. In October this may be washed and cleaned from the weeds which will have grown in it during the year. Then so soon as the fish are ready to spawn they will ascend from the ponds into the raceway seeking a place to nest. Then they are ready to be taken out and the spawn expressed. At the entrance of the raceway there should be a groove to receive a frame on which is tacked a net of coarse bagging about eight or ten feet long. One corner of this bag should be narrowed and tied with a string, like the mouth of a meal-sack. The race should be covered over in spawning time, as the fish will come under the cover better and are not so likely to be frightened at any one passing that way. If there are fifteen hundred or two thousand fish in the pond the net may be used every day in the height of the season, and when the fish become scarce, once in two or three days.

Indications of spawning having been observed, the covers are put on the races, and being satisfied by sly peeps through the cracks of the covers that there are fish in the raceway, the net is gathered up in one hand and the frame held in the other, in such a position as to be put in the groove as quickly as possible so as to let none of the fish escape from the race. Go quietly to the spot, and do not walk down the raceway to get to it, else you will frighten the fish ; but approach from one side and put the net in the groove as quickly as you can. The water running down will swell the net out to its full length. The covers may then be removed, and with a switch you may frighten the fish down from the head of the raceway into the net. As soon as they are all in, the frame may be lifted out of the water, and the fish are then enclosed in the bag. A tub of water should be previously brought near the spot, and the end of the net can be lifted into the tub and untied, when the fish will all fall into the tub without trouble. Coarse cloth is better for the purpose than netting, as it can be more easily tacked to the frame, does not hurt the fish so much, and lasts longer ; besides this, the water swells it out and holds it open for the fish to run in ; a net would not bag out so well, and the fish not seeing you through the cloth as they would through an open mesh are not scared, and do not try much to run back up the race. It must be remembered in this and all subsequent handlings of the fish, that if the outer skin of a Trout is broken or abraded by the hand or by contact with any hard substance, it will, in nineteen cases out of twenty, cause the fish to die. A white fungus appears on it first where the skin is broken ; this fungus spreads over the fish until it is sometimes half covered with it before it dies. We speak of the covering of the Trout as "skin," because it feels like it and looks like it, although in reality a Trout is covered with minute scales. They will get over a deep and clear cut much more quickly than over a bruise where the cuticle or skin only is broken.

The fish being now in the tub must be taken to the hatching-house as quickly as possible. There are probably in the tub some fifteen or twenty fish, and all the operations must be conducted as quickly as possible so that the fish will not die in the small quantity of water to which they are confined. So long as the fish lie quiet in the bottom of the tub there is sufficient air in the water to sustain them, but so soon as they begin to jump to the surface of the water and try to leap from the tub, it is a sign that the air is nearly exhausted and the water should be renewed. They will also open their mouth wide, just as a person would when gasping for air. The question has sometimes been asked how long a Trout would live out of water; the answer is, about as long as a man would live under the water. Trout will die in a tub out of which the oxygen has been exhausted by their breathing, more quickly than they would die in a cloudy day if out of the water entirely.

A fire may be made in the hatching-house to warm your fingers, which will probably get cool while engaged in this operation. A six-quart milk-pan is to be provided, filled half full of water, if you have many fish, and also another tub of water, into which to put the fish after they are deprived of their spawn. Select a male fish, and holding him over the milk-pan, with his belly under the surface, rub it gently with the fore finger and thumb, from the pectoral fins to the vent. A little experience will show how this is to be done. If the fish is ripe, a few drops of pearly or milk-colored milt is forcibly expressed into the water. If the milt is not of this color, it shows that the milt is not good, and another male must be taken and treated in a similar manner. Go on until a good one is found. But if the first one is not good, it is probable that the second one will be all right. Then take a female fish, and pressing it in the same way, the eggs will be exuded.

The female must be pressed more slowly and oftener than the male. If the eggs are not ripe, by passing the hand lightly over the belly, you will feel them beneath, hard, like shot. In that case put the fish back into the pond, for the eggs to ripen. When the eggs are ripe, the belly will be soft and flabby, and the eggs beneath the skin feel loose and change their positions at the touch. So loose are they, that by holding the fish in a horizontal position, and then moving it up and down, the eggs will change their position in the womb, and fall downwards or upwards, as you hold the fish head downward or upward. The operation must be continued,—first a male and then a female, until the fish are exhausted. The water in the pan may, at intervals, be gently stirred with the tail of the fish you are holding in your hand; this is to change the position of the eggs, so as to be sure that all the eggs come in contact with the milt, and the tail of the fish is better than anything else to stir with, besides being ready in your hand. The pan should then be set in one of the hatching troughs through which the water is running; this will keep the eggs up to the proper temperature, and prevent a sudden change when they are transferred to the trough. The eggs will now agglutinate or stick to the pan, and to each other, for a little while.

Now then, having first put the fish back into the ponds, while you are waiting for the eggs to separate, we will say a few more words about handling the fish. A Trout, it is well known, may be tickled under the belly, and rather seems to like it, and will lie quiet in your hand while you are doing it. By putting the hand gently in the water, moving it cautiously around the fish, and gently lifting, he may be raised high and dry, and lie quietly without a struggle. But this mesmeric operation takes too long to try in spawning, and had better be left out of the question. There is a way, however, of grasping a Trout firmly, but gently. So fast that he cannot squirm, and yet not hard enough to break the skin. This art, in its perfection, is one which only great masters attain ; neither this, nor the subsequent manipulation, can be well done until after long practice. The novice will probably kill one-quarter of all the Trout he attempts to handle, and if, after years of experience, he loses only three or four out of every hundred, he may think himself pretty well advanced in the art. The fish must be grasped by the head (if you are right-handed), with the right hand, and by the tail, or rather the lower part of the body, with the other hand, and held over the pan with the belly in the water. As soon as the fish is quiet, the right hand may be gently slipped down from the head, and the fore finger and thumb used to press the belly. The fish still being held by the tail, and lower part in the left hand, and partly supported, perhaps, by the sleeve of the coat, or by the bare arm, and the remaining fingers of the right hand. Every one will have a way in which he can do this best, and will find it out after a few trials. If the fish is large and struggles violently, the usual direction given in the books, is to let an assistant hold the tail. We counsel you, if the fish struggles violently, whether it be large or small, to drop it back into the tub, manipulate another, and after a few minutes, try it again ; it will lie quiet after a while, especially if it is a small fish. If you attempt to hold a fish, which struggles violently, you will be very apt to kill it. If, in addition to your own two hands, you get the two hands of an assistant, on the struggling fish, there is not much chance of saving him alive. Here is a better way. File the barb off of a No. 4 hook, then tie it with three feet of line to a pliant switch, three feet long. Hook your fish on this, through the jaw, and, holding it in a tub of water, let it struggle until it is exhausted. Then the hook can be slipped out, no injury having been done to the Trout, and it can be handled without any difficulty.

The large fish are harder to handle, struggle more violently, and are more apt to be killed than the smaller ones
and do not average so many eggs, although now and then
one will have a very large number. Therefore, we consider
that the best fish for breeders, when the operation is conducted by hand, are those weighing from one-quarter of a
pound to one pound.

While the eggs are standing in the pan, at intervals o
three minutes, give the pan a gentle shake—just one shake.
While the eggs remain this way in the pan, the milt is
coming into contact with them and impregnating them ;
the object of the shake is to change the position of the
eggs so as to get them all fairly exposed to the influence
of the milt. It takes very little milt to impregnat
a large number of eggs. Enough to slightly tinge the
water, will impregnate a pan full of eggs. But, in practice, we generally take all the milt we can get out of
the haul. It is sometimes our custom also to put the male
fish, whose milt has been exhausted, into a pond by themselves, to keep them from running up into the race again,
and troubling the females. This is a very good plan, if
you have plenty of ponds and plenty of fish. If you have
but a small number of males, compared with the number
of females, put them back again into the pond, as they will
probably have a second and third edition of milt.

Twenty to twenty-five minutes having now elapsed since
the pan of eggs was set into the trough, gently tip up the
pan. If the eggs are loose from the pan, and roll separately
as you move it, we are ready for subsequent operations ; if
not yet loose, let them be a while longer.

We know that the semen of the male is full of animalcules. These will live for ten or fifteen minutes. There
is a hole for the reception of these animalcules in each
egg. The egg always sinks into the water with this hole
at the top. It receives one of the animacules only by this

opening, and then closes.   It seems to be a special arrangement of Providence that the eggs shall agglutinate—stick
fast to each other and to everything they touch—so that
they shall not float away until they are impregnated and
the Trout has had time to cover them.   In the eggs of
other fish, such as Bass and Perch, the same arrangement
is seen ; only they stick fast the moment they touch anything, and stay there until hatched out, while that which
fastens the eggs of the Trout dissolves as soon as the
mother has had time to cover them.

The eggs will now be loose and lying on the bottom of
the pan.   Pour off the dirty water until only sufficient is
left to cover the eggs.   If this is done very gently, the
eggs, although very light, will remain at the bottom of the
pan, as they are somewhat heavier than water ; then sink
the pan into the water, at the same time tipping the pan,
as described in the chapter on Eggs, and take the pan half
full of water.   The influx of water will wash the eggs
around somewhat, and dilute the dirty water remaining in
the pan.   This is to be poured off, as before, and the operation repeated, until the water in the pan looks perfectly
clear.   There will be some dirt and droppings of the
Trout still left, which can be carefully picked out with the
nippers.   If an egg should happen to be broken, while
being taken from the Trout, every vestige of it should be
carefully removed, as the slimy, sticky contents will get on
the other eggs and kill them.   The eggs are now ready to
be placed in the trough, and having previously raised the
water in the nest to which you wish to transfer them, by
placing a strip across, proceed as described in the chapter
on Eggs.

From the above description, it will be seen that a few
lessons in artificial impregnation, from an experienced
hand, will probably save the beginner much time and
money.   A written description of the process, however

good, can never take the place of verbal instruction ;
partly because it never conveys exactly the same idea to
all, partly because seeing a thing is better than hearing
about it, and most of all, because a written description is
a general one, and hardly ever tells of the minutiæ and va-
riations which constantly occur in practice. As an exam-
ple of this, it has been urged, all through this book, that
in moving the eggs the beginner should not touch them
with the feather, but should move the water over them,
with the feather, so that the eggs should follow the current
thus created ; also that he should be very careful, in removing the dead eggs, not to touch the others with the
nippers. But, in our establishment, we constantly move
the *eggs* with the feather, and push to one side the sound
eggs with the nippers, in order to get at the dead ones.
The reason simply is, that long practice has given the
knack of doing it, without injury to the eggs, and a tyro
could not do it.

TAKING SPAWN BY AINSWORTH'S SCREENS.—Mr. Stephen
H. Ainsworth, the pioneer of fish-farming in this country,
last year invented an apparatus, which is, we believe, des-
tined to work a revolution in the whole science of fish-
breeding. It has not yet been tried sufficiently to pass a
final judgment upon its merits. But so far as it has been
tried, it has answered a good purpose, and generally met
with success. From a perusal of the preceding method of
taking the spawn by hand, it will be seen that many difficulties attend the operation. In the first place, it requires
long practice and a natural gift besides, in order to succeed in
it well. Then, many of the fish are inevitably lost through
being handled ; and the operation is a very inconvenient one.
Requiring, as it does, to be performed in the depth of winter, when the ground is covered with snow, and water on
one's hands and clothes soon becomes ice, it is a cold and

6

disagreeable job. The plan which Mr. Ainsworth has invented obviates, in a great degree, these objections. As his apparatus is not patented, but generously offered free to the public, we describe it here. The difficulty in getting eggs impregnated naturally has been that the Trout would eat so many of them, when laid, and that being scattered about, on and under the stones and gravel, they could not easily be collected, and many, in any event, would be lost. The problem which Mr. Ainsworth had to solve was, then, to keep the Trout from eating the eggs after they were laid, and to devise a plan by which they could be easily collected. This he accomplished by laying in the raceway two wire screens. The lower one of such fine wire that the eggs will not pass through ; that is, of about ten or fourteen threads to the inch. This wire is attached to a frame, made of inch stuff, and another inch strip nailed above it. Another frame is provided, of the same width and length, but the sides of which are from three to five inches deep ; upon this a coarse screen, of three or four wires to the inch, is fastened. The fine screen is first laid in the race, which being made of proper width, it fills, and the coarse screen is laid over it, with the wire side down. And there is a space, between the two screens, of one inch, protected from invasion on the top and bottom by the wire screens, and on the sides by the inch strip, on every side of the small screen. The top screen, which has sides three or four inches deep, is then to be filled with coarse gravel (so coarse that it will not pass through the meshes), to the depth of two inches. This gravel will overcome the buoyancy of the wooden frames, and cause them to sink into the water. Now the screens are ready for use. Let us see how they operate. A Trout comes along, and finds the gravel. She sees no screens—only some nice gravel for nest-building, in what appears to be a shallow box. Suspecting no evil, she proceeds to make her nest, and in the process of "slirting out" gravel with her tail, she moves it away from the meshes of the coarse screen, and leaves the bottom of her nest an open network. On this she emits her eggs, which are at the same time fecundated by the emission of the milt of the male Trout lying by her side. The eggs fall down into the nest, but pass through the coarse wire screen, and are caught by the fine meshes of the lower screen. There they are safe. The Trout covers up the hole as usual. The hangers-on find no eggs to devour, and go their way. Another Trout may make her nest in the same place, without disturbing the eggs already laid, safe in their resting-place; and when a pleasant day comes, and you feel in the humor, you may take a pan of water, and taking off the upper screen, gradually lift up the lower screen, brushing the eggs to one corner with a feather, and tip them all at last into your pan, without having exposed a single egg to the atmosphere, without any trouble in handling the fish, and without any loss of the breeders. These screens may be made as wide as your raceway, if it is not over two or three feet, and of a square shape. If your raceway is four feet wide, it is better to have your screens each two feet square, as this size is most convenient to handle when they are filled with gravel. Enough of them can be placed in the raceway to fill its whole length. One thing requires to be noted here. It takes a much larger raceway, for this process of natural impregnation, than it does when the eggs are impregnated artificially. In the latter case you need only room enough to make one Trout after another *believe* she is going to lay her eggs; in the other you must have room enough for her to carry out her intentions.

Much discussion has taken place among fish-breeders and others interested in the art, as to the comparative value of the two methods, aside from the manual labor and loss of fish involved. That is, by which of the two methods the

most eggs are impregnated, and which are the most healthy
or will produce the best fish.   In answer to the first question, we answer that a skillful workman will impregnate
more by hand than are impregnated in the natural way.
But there are very few such skillful workmen.   As the
eggs of a Trout are not all ripe at the same time, the unripe
ones are lost in hand impregnation ; but then, some are
lost also in the screens, by falling upon the gravel and
not passing through the wires, and also by being bruised
in passing through.   To the second question, we answer,
that the fish hatched from those rightly impregnated
by hand, are just as strong, grow as fast, and live as
long as the others.   Still, the advantages of Ainsworth's
screens, for taking impregnated eggs, especially to a novice,
are so great, as to preclude from him all question of artificial manipulation.

CHAPTER VIII.

GENERAL REMARKS.

STOCKING PONDS.—The question is often asked by begin-
ners, with what shall I commence fish-farming? Shall I
buy the eggs and try to raise them, and wait three years
for full-grown fish, or shall I buy adult fish, and from them
take eggs, etc.? The answer to this question depends upon
two circumstances. First, how much money you have;
and second, how long you wish to wait. It is much cheaper
to buy the eggs than the adult fish; but then you will have
to wait two or three years before you have any breeders.
My own advice would be to try a few thousand eggs, and
also a few hundred two-year old fish. At the present
prices (1869–70), ten thousand eggs would cost eighty dollars, and two hundred two-year olds would cost about fifty
dollars. Two hundred two year-olds would probably
give about twenty thousand eggs. If you take this advice,
you will have eggs to experiment with the first year.
With care, you will hatch out more or less, but in any case
your experience will be invaluable to you for the next year,
and you will have a stock of breeders, to furnish eggs, as
you want them.

STOCKING STREAMS.—Persons who own Trout-streams
would very often like to have them re-stocked, and some
make feeble attempts to do it, by putting in a few thousand
young fish. This would re-stock a small stream, if it was
done every year, for some years. But it is folly to suppose
that a stream which has been fished for years, and thousands
taken from it every year, can be re-stocked *quickly* by putting

in a few hundred, or even a few thousand young. If an exhausted Trout-stream had ten thousand young put into it, every year, for three years, it would then stand the strain of moderate fishing to the end of the world. If you attempt to stock your streams at all, don't do it half-way. Remember that the less fish you put in, the longer you will have to wait. It is much easier to stock a stream than to raise fish in ponds; because the young fish will take care of themselves better than any one can take care of them; and if they are protected from danger, until they are forty-five days old, they are then tolerably able to take care of themselves. In stocking a stream, the young fish should be taken to its head-waters, or put into the springs, or little rivulets, which empty into it. As they grow larger, they will gradually settle down stream, and run up again to the head-waters in winter to spawn.

When putting fish into a stream, do not put them suddenly into water much warmer than that of the vessel in which they are brought. They will not so likely be injured by putting them into colder water; but try to avoid all sudden changes, and gradually raise or lower the temperature of the water in which you bring them, until it is even with that of the stream in which they are to be placed.

WILL IT PAY?—This question has been asked of us more often, perhaps, than any other, and is more difficult to answer. The same energy and tact, which will make any other business pay, will make this pay. Many men start in the dry-goods business; not many succeed. And yet the dry-goods business pays. While Trout are selling at a dollar per pound, the business certainly must pay. Even if they come down to twenty-five cents per pound, the profit must still be large. But all persons do not have the natural ability required for this business, and would not succeed in it. Not all men can be good farmers, or lawyers, or pisciculturists. The question of profit depends much upon the cheapness of food, and here is another point where farmers have the advantage in this business, as they have the bulk of the food ready to their hands, and fattening Trout will pay better than fattening pigs.

If much competition arises, even then it will be the same as in any other business. He, who by economy in practice and the use of improved methods, can sell his fish the cheapest, will make the money. If the time should come in the future, when, on account of competition, but little money can be made at the business, then competition will naturally be reduced, and the price will rise to a pay-ing figure. People will not raise Trout for sale, unless they can make money by it, and so long as there is any de-mand for Trout, somebody will have the chance.

ADVICE TO THOSE STARTING IN THE BUSINESS FOR PROFIT. —Remember that it will be two years before you can hope to sell any fish for table use; so do not enter into it unless you have some means of support for that time. The persons who have the most natural advantages for this business are those farmers, who have springs or cold streams on their farm—now almost useless—but which may be turned to advantage in raising fish. They, depending on their farm meanwhile for support, can yet give time and attention to the experiment, and engage in it altogether if it succeeds, or abandon it, without serious loss, if they fail. It is peculiarly adapted to them, also, because it demands most attention in the winter, when they have least to do on the farm. Meanwhile, until fish-farming becomes rather less of an experiment, and more of an exact science, it would be well for impecunious young men, seeking for-tunes, to leave the business to capitalists and corporations. To those who wish to raise fish for their own table use, or to afford sport in angling, to themselves and friends, we would say that we can think of no way in which a little time and money can be so well laid out as in Trout-raising.

In the preceding chapters, it may appear to the casual reader that we have given more details than were necessary; but we are sure that every beginner of fish-culture will think differently. We are also so used to writing letters and answering questions on the subject, that in our book we have said, " you must," and " you should," a great deal more often than was required. However, if we have got the facts right, nobody, we suppose, will complain of our taking liberties with the second person singular.

PATENT ROLLER SPAWNING BOX.

My partner, Mr. A. S. Collins, has recently invented an improvement in the method of taking naturally impregnated spawn. This improvement removes many of the objections to Ainsworth's screens, as formerly used, and is, in my opinion, of very great value. We have had one in operation at our ponds the whole of this season; it has worked to our perfect satisfaction, and next year (1870) we shall put them down in every race we have. The principle used is that of the Ainsworth's screens, and the improvement consists in the method of collecting the eggs.

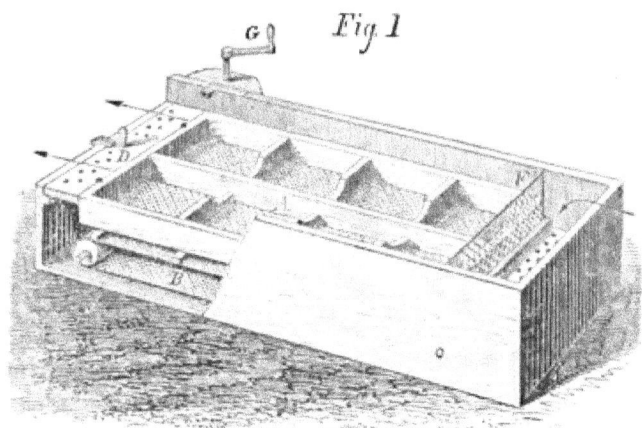

Fig 1

Figure 1, represents a small spawning box with a portion of the side removed. Figure 2, (on the next page), is an enlarged view of the front of the same box. At A, is seen a double row of coarse wire screens, eight in number, (three meshes to the inch).

These screens are to be filled with coarse gravel,
and the eggs pass through as in Ainsworth's screens.   Under these is an endless apron of fine wire cloth, (B), passing over rollers at the two ends of the box.   This apron of
wire cloth is about one inch beneath the upper screen, and
is kept in its place and prevented from sagging by small
cross-bars, (two of which are seen in the cut), correspond-
ing to the divisions of the upper screen and running in
grooves in the sides of the box.   These cross-bars also keep

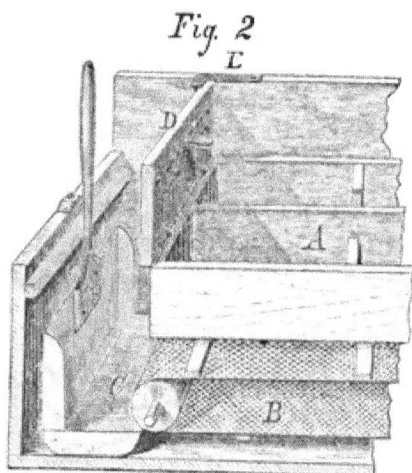

Fig. 2

the eggs from being carried down by the current.   The
front roller can be turned by the handle seen at G.   As
the roller is turned forward it moves the screen with it,
and, of course, the eggs as they come to the edge of the
roller will fall off.   The pan, C, (fig. 2) is placed in front
of the roller, receives the eggs as they fall, and the operation is complete.   The box need not be more than two feet
deep, and may be made only eighteen inches deep; it is
set directly in the raceway, and intended to fill it completely.   The water enters in the direction of the arrows,
and may either enter with a fall over the top of the box,
as seen in figure 1, or the top of the box may be cut down

until the water will enter on the level at which it is to stand over the screens.

F, (fig. 1), is a screen intended to prevent any fish getting to the lower screen, either from within or from without, and may extend to the bottom of the box.

D, is a screen for the same purpose at the front of the box. When the eggs are to be taken, the screen is raised on hinges to an upright position, as seen in fig. 2. This confines the fish which may happen to be in the race, and none of them can get below. The pan is then lowered to its position, the roller turned and the eggs taken. When the operation is finished the screen, D, is again lowered, the button turned, and the work is done. If the box is wide—say four feet—it is more convenient to have the pan made in two or three sections, inserted in a light frame, as the eggs can be more easily carried in and poured out of a shorter pan.

The box can be of any length from four feet to sixty feet, or even longer, and of any width, from two feet to six or eight. If it is made very wide, an additional longitudinal support must be provided for the revolving screen. We recommend the following dimensions for Speckled Trout races: Two feet deep, two feet wide, and from ten to twenty feet long, or four feet wide and twenty to forty feet long. The upper screens may be made in convenient sections, the whole width of the box and six or eight feet long.

A few of the advantages of the plan are as follows:— Compare a double row of two feet square, Ainsworth screens, forty feet long and four feet wide,(such as we have now in use), with one of our spawning boxes of the same dimensions.

1st. By the old way, it would take two men a good half day to remove the screens singly, feather off the eggs in a careful manner and return each screen to its proper place.

It would take the new spawning box about fifteen minutes to do the same work, with one man.

2d. The weight of the gravel which has to be lifted in the old way, amounts to many tons in the course of the season.

In the new box, the gravel is not lifted at all.

3d. By the old way the operator's hands must, of necessity be more or less wet during the whole operation. Now, as the Trout, &c., spawn during the winter season, it may be imagined how pleasant this feels when the thermometer is anywhere below the freezing point.

By the new way the hands are not made wet, and may be kept comfortably gloved.

4th. By the old way, more or less of the eggs are lost by careless feathering, exposing the eggs to the freezing atmosphere, clumsiness in handling the screens by cold fingers, &c., &c.

By the new way, every egg is saved.

5th. By the old method every fish is driven out of the race when the eggs are taken. Some of them will not return, but will seek a spawning place in the pond, and many eggs will be unavoidably lost.

By the new way the fish are not driven from the race. And, as the boxes are always covered during the season, the fish will not even be disturbed. In fact they may spawn *while the eggs are being taken*, and yet not a single egg be lost.

This spawning box answers for securing the naturally impregnated eggs of Salmon, Salmon Trout, Speckled Brook-Trout, Whitefish, Shad, &c., &c.

The screens, F and D, are so made that while a full current is permitted to flow over the upper screens, (A), only a gentle current can flow through the under part of the box. This current is meant to be so regulated, that when the pan, C, is placed about an inch from the turning roller, all the small stones which the Trout may whip through the upper screen will fall short of the pan, the eggs being lighter will be carried by the current into the pan, while a great part of the dirt, &c., which may collect on the lower screen, will be carried up over the pan and entirely out of the box. The under screen may be made of tarred muslin or mosquito netting. But wire cloth (8 or 10 meshes to the inch), keeps much the cleanest, and we are inclined to think it best for the purpose.

A brush may also be placed under the forward roller; so that every time the roller is turned the screen will be perfectly cleaned.

The box looks, at first sight, somewhat complicated in structure, but is, in reality, perfectly simple. Any one who has a knack of using tools can make such a box, which would answer the purpose perfectly. The cost is very little more than that of the old screens, (same area). The extra expense for two rollers, two hinges and handle, being, perhaps, two or three dollars.

We are now ready to sell, for a moderate price the right to make and use such boxes, and will give to purchasers full description, &c. Address, SETH GREEN & COLLINS Caledonia, N. Y.

APPENDIX.

FISH FARM AT CALEDONIA, N. Y.

This celebrated fish-farm is three quarters of a mile from Caledonia, N. Y. Caledonia is a small village, about seventeen miles south-west of Rochester, and seven miles west of Avon Springs. Both the New York Central and Erie railroads have stations at Caledonia. The Rochester and State Line railroad, recently planned, will have a station at the village of Mumford, which is one-quarter of a mile from us. Caledonia is noted for its creek, which rises entirely from springs, is fed along its whole course by springs in its beds, and at our fish-farm, which is about three-quarters of a mile from the source, it runs about eighty barrels of water per second, 4,800 per minute, or something over 200,000,000 of gallons in twenty-four hours. Quite a respectable quantity of water, and the whole of it available for our ponds, if we wish to use it. The ground in the neighborhood being very level, no surface-drainage of any account washes into the creek, and the water looks pure as crystal. It is, in reality, slightly tinctured with lime and sulphur; but must agree with the fish, as the creek has always been noted for its Trout, and still abounds in them. This fish-farm is now owned by Seth Green, A. S. Collins and S. M. Spencer. It embraces within its limits one-half mile of the creek. The ponds, etc., being arranged for a special pur-pose, are not laid out according to the plan given in this work. Besides this, some of them were laid out when we first commenced fish-culture and did not have the experience which we have now gained ; still as the supply of water is so large, they serve a very good purpose, even in their present state. Although the intention has been to raise fish for market, yet the reputation of the senior member of the firm, as a fish-culturist, and the publicity given to our farm, through the papers, has caused such a demand upon us for eggs, young fry, and adult fish, for stocking ponds, that it taxes all our resources to supply this demand, without selling a pound of fish for eating. As we take the greatest of care in collecting the eggs and packing them, and from inspection with a miscroscope, before sending them away, are sure that a good per centage are impregnated, our eggs are sure to hatch out, with ordinary care and ability. Our experience and care, together with the reputation which some years of fish-farming has given us, are guaranties that everything obtained from us will be of the best. As before said, we do not guarantee success to every one; that depends upon the individual who makes the trial. But we do guarantee to give him the best of materials to work with. And he will succeed, unless he is careless or incompetent. " That which has been done once, can be done again."

Our prices are as follows (1869-70), for Trout ova, by the single thousand, ten dollars. For five thousand, or over, eight dollars per thousand. We send them from December 1st to March 1st. They will be sent by express at our risk, and can be sent anywhere in the United States, to Canada, England, or France. To the Southern States, they can be sent with safety only in the coldest winter weather. Young Trout fry, one inch long, we sell at the ponds, or deliver at our nearest express station, for thirty dollars per thousand. We sell them from Feb. 1st to May 1st. In cool weather, one thousand can be carried safely a two days' journey, in four gallons of water. The best time to send the young fry is four or five days before the umbilical sac is off the belly; then they do not suffer for lack of food, and do not require so much water. One year old Trout, we sell at about twelve dollars per hundred. Two-year old, about twenty-five dollars per hundred: the price depending somewhat upon the size of the fish. We pick out an equal number of males and females, in good condition, and charge a little more than the market price for dead fish. Eggs we can send safely, by express, any distance; but fish have to be sent in cans, or barrels, or tanks of some kind, and the employes of the Express companies will not always bother themselves, even to the extent of keeping the cans or barrels right side up. Fish can be carried alive, in cold weather, with much less trouble than in warm weather, but can be sent any time, if some one goes along to take care of them. If we send a man with them, we charge for his time and travelling expenses, and send the fish at our risk. Otherwise, we sell the fish at buyer's risk.

We give lessons in the art of Trout-culture at ten dollars per day (Board can be obtained at the hotel, one-quarter of a mile from us.) These lessons consist of instruction in the method of handling the fish, and all the minutiæ of the art, and are best given from January to May, because then all our operations are in progress; but will be given as well as possible, at any time. We also lay out ponds and stock streams for any one desiring our services. Our charge, for personal attendance on such business, is ten dollars per day, and expenses.

A detailed description of our farm has been given so often in the papers, that it is unnecessary here, and besides would not be of much use to the beginner, as our arrangements are made with reference to getting the greatest possible quantity of eggs and young fish. We sell a few of the largest and most unwieldy breeders each year. But, if our business in stocking ponds increases, in the same ratio as it has done in past years, we shall have more than we can manage. A few items, in regard to the ponds, may be of interest. We have one pond, seventy feet by ten; one, thirty by ten; two, thirty by twenty; one, thirty by twenty-five; besides many other smaller ponds. These are not very alarming proportions; but the reader will see that we practice what we preach, and use small ponds with large supply of water. We use about ten thousand of our adult fish as breeders. We run so much water through the ponds (never using the same water twice) that we could fill them, if necessary, with fish, so that the bottom would be entirely hidden from sight, and the fish would do well. Our hatching-house has a capacity for starting two million of eggs; that is, it will hatch out that number of eggs; but there would be too many, after hatching, for the amount of water. If we wished to keep them in the troughs, two or three months after hatching, we would not put in more than two hundred thousand. The grounds are being improved, year by year, and are getting to be quite a resort for the summer visitors in the neighborhood. The water in our stream never freezes; does not fall below forty-five degrees in winter, nor rise above sixty in the hottest summer weather. The volume of water is almost uniform, being very little affected by drouth or rain, and does not vary through the year more than four or five inches. Altogether, it is one of the best places we have ever seen for the purpose, and we doubt if it can be equalled.

Most of our States have laws relating to the protection of fish, similar in tone to that of New York. We give that portion of it relating to Trout, thinking that it may be of use to the breeder.

GAME LAW OF NEW YORK,

RELATING TO TROUT.

SEC. 13.—Any person trespassing upon lands owned or occupied by another, for the purpose of shooting, hunting or fishing thereon, after public notice by such owner or occupant as provided in the following section, shall be deemed guilty of trespass, and shall be liable to such owner or occupant in exemplary damage for each offense not exceeding twenty-five dollars, and shall also be liable to the owner or occupant for the value of the game killed or taken.

SEC. 14.—The notice referred to in the preceding section shall be given by publishing an advertisement particularly describing such land, and forbidding such trespass in the official papers of the county, or a paper published in a town where such lands are situated, for the period of three weeks, and in the months of April and May in each year, by sign-boards, at least one foot square, to be put up and maintained in not less than two conspicuous places on the premises; such notices to be signed by or have appended thereto the name of the owner or occupant.

SEC. 15.—No person shall place in any fresh water stream, lake or pond, without the consent of the owner, or in shore waters and estuaries with rivers debouching into them, any lime or other deleterious substance, with the intent to injure fish; or any drug or medicated bait, with intent thereby to poison or catch fish; nor place in any pond or lake stocked with and inhabited by Trout or Black Bass, any drug or other deleterious substance, with intent to destroy such Trout or Bass; nor place in any fresh water, pond or stream stocked with Brook-Trout any Pike, Pickerel, Black Bass or Rock Bass, or other piscivorous fish (Salmon excepted) without the consent of the owner or owners of such lands upon which such pond or stream is situated. Any person violating the provisions of this section shall be deemed guilty of a misdemeanor, and shall, in addition thereto, and in addition to any damage he may have done, be liable to a penalty of one hundred dollars.

SEC. 17.—No person shall at any time, with intent so to do, catch any speckled Brook-Trout, or any speckled river Trout, with any device save only with a hook and line, and no person shall catch any such Trout, or have any such Trout in his or her possession save only during the months of April, May, June, July and August, under a penalty of five dollars for each Trout so caught or had in his possession; but this section shall not prevent any person or corporation from catching Trout in waters owned by them, or upon their premises to stock other waters, in any manner or at any time. But the counties of Kings, Queens and Suffolk shall be excepted from the provisions of the above section, so far as to allow the taking or catching of Trout in the counties last named during the month of March.

SEC. 18.—Any person, or persons, or company engaged in the increase of Brook-Trout by artificial process (known as fish-culture), may take from their own ponds, in any way, and cause to be transported, and may sell Brook-Trout and the spawn of Brook-Trout at any time, and common carriers may transport them, and dealers may sell them, on condition that the packages thereof so transported are accompanied by a certificate of a Justice of the Peace, certifying that such Trout are sent by the owner or owners, or agent of parties so engaged in fish-culture. And such persons or company may take, in any way, at any time upon the premises of any person, under permission of the owner thereof, Brook-Trout to be kept and used as Brook-Trout for artificial propagation only, and for no other purpose.

TRANSPORTATION OF LIVE FISH.

Many expensive tanks have been constructed for transporting fish alive, answering the purpose more or less perfectly. We give here a simple and inexpensive method: Take a barrel or cask, washed until it is clean and sweet. Fit a cover to it tightly to prevent the water splashing over while in the cars or wagon. A piece of canvass tied over the top, answers every purpose. A hole one inch in diameter may be made in the middle of the cover. Fill with water within six inches of the top, as the agitation of the water on the journey helps to aerate it. Tie some ice in a piece of flannel and fasten it to the side of the cask near the top so that it shall not swing about and bruise the fish, and the cold drip from the ice will sink to the bottom. If the journey is to be a prolonged one, fit the nozzle of a common bellows with a tin tube long enough to reach to the bottom of the cask, and by blowing a little now and then the fish can be carried thousands of miles. We do not give this as the best plan, but as a cheap and inexpensive method answering a very good purpose. The best apparatus would be a metal tank of some kind with double walls, permanent ice chamber in the middle, and automatic air-pump.

YOUNG TROUT CAUGHT IN WEBS.

Since finishing the body of this work I have been engaged in hatching out Whitefish and have discovered something which I wish to note here. There is a small worm which is a favorite food of Trout and many other kinds of fish. This worm is one of the greatest enemies which the young fry have. It spins a web in the water to catch young fish, just as a spider does on land to catch flies. I have seen them make the web and take the fish. The web is as perfect as that of the spider and as much mechanical ingenuity is displayed in its construction. It is made as quickly and in the same way as a spider's, by fastening the threads at different points and going back and forth until the web is finished. The threads are not strong enough to hold the young Trout after the umbilical sac is absorbed, but the web will stick to the fins and get wound around the head and gills and soon kills the fish. I have often seen it on the young Trout and it has been a great mystery and caused me many hours, days and weeks of study to find out what was wound around the heads and fins of my young Trout and killed them. I did not find out until lately, while watching recently hatched Whitefish. These are much smaller than the Trout when they begin to swim, and they were caught and held by the web. I found ten small Whitefish caught in one web in one night. This web was spun in a little Whitefish preserve, into which I had put one hundred young fish. The threads spun by this worm seem to be much finer than the common spider's web, and they are not visible in the water until the sediment collects upon them. They can then be seen very plainly. These webs cannot be spun where there is much current and can be easily seen in still water by a close observer.

www.ingramcontent.com/pod-product-compliance
Lightning Source LLC
Chambersburg PA
CBHW020256090426
42735CB00009B/1100